Murdo Wright

Native silica, a treatise upon a series of specimens of quartz, rock crystal, chalcedony, agates and jaspers, as well as other earthy and metalliferous minerals

Murdo Wright

Native silica, a treatise upon a series of specimens of quartz, rock crystal, chalcedony, agates and jaspers, as well as other earthy and metalliferous minerals

ISBN/EAN: 9783742814807

Manufactured in Europe, USA, Canada, Australia, Japa

Cover: Foto ©Klaus-Uwe Gerhardt /pixelio.de

Manufactured and distributed by brebook publishing software (www.brebook.com)

Murdo Wright

Native silica, a treatise upon a series of specimens of quartz, rock crystal, chalcedony, agates and jaspers, as well as other earthy and metalliferous minerals

PREFACE.

SPECIMENS of Silica, under the names of Agate, Sard or Carnelian, Jasper, and other ornamental semi-precious stones, were held in the highest estimation from the remotest ages on account of their beauty and imperishable nature, which latter characteristic induced the ancients to use them as the media for handing down to posterity the deeds of their heroes and statesmen, engraved in the form of Camei and Intaglie.

One of the earliest stones of which there is a reliable account is the ONYX, mentioned in the Book of Genesis.

Almost all the Silicas are, in fact, mentioned by ancient authors, the nomenclature often varying, however, from that of the present day. Some of them occupied places in the Breast-plate and Ephod of the High Priest.

Theophrastus wrote his valuable Book on Stones (Περί Λίθων) about 315 years B.C., and Scaurus, so it is recorded, was the first to collect valuable stones, which included at that period the Marbles as well as Silicas and other Minerals, about the year 60 B.C.

It is to Pliny the Second, frequently called Pliny the Elder, that we are indebted for putting together, in something approaching complete and compendious form, all the stores of information with regard to the Mineral Kingdom extant in the first century, A.D. From his justly celebrated and ency-clopedic work, the *Historia Naturalis*, a true notion may also be formed of the high estimation in which ornamental stones were held by the Greeks and Romans.

Reference has been made in the present volume to his work, in all cases where it was available.

The taste for collecting ornamental Minerals, begun in those early ages, has never been lost, but has extended itself through the East, to Greece, Italy, France, and other parts of Europe.

In modern days several valuable collections of Agates have been made, notably that by the late Professor V. Nöggerath, which is exhibited at the Museum of Practical Geology. The "Birdwood" series of Indian Agates collected by Dr. (now Sir George) Birdwood, in Bombay, which was merged into the great collection made by the late Mr. Arthur Wells, of Nottingham, was dispersed at Christie's, and from that sale the examples in this Collection were procured. The collection made by the distinguished collector, Mr. Alfred Morrison, comprises a complete cabinet full of most beautiful examples ; and, above all, we should notice the collection made by Professor John Ruskin, rendered doubly valuable by the fact that it contains the examples he has so ably described.

The formation of the "Derby" Collection, described in the following pages, commenced about the year 1870.

The late Earl of Derby was particularly struck with the exquisite Agate No. 224, the formation of which, in most minute regular concentric zones, embedded in crypto-crystalline Silica, is truly marvellous. His Lordship purchased it together with a few others, which formed the foundation and nucleus of the Collection, since which it has steadily increased, more particularly during the last six or seven years.

All specimens of sufficient merit arriving in London were submitted to Lord Derby's inspection, and a careful selection made, needless to say, with the most refined taste, resulted in the unique, and, it may be fairly claimed, most beautiful collection of cut specimens of Agates and allied minerals ever made.

It is not a *general* Collection of Minerals,—its scope being restricted to Silica, the important group which includes, not only Quartz, the commonest of all Minerals on this Earth, in all its forms, but the rarer and more beautiful varieties used in ornamental and architectural art.

The Collection, bequeathed by Lord Derby to the Liverpool Free Museum, numbers seven hundred and eighty-two specimens, of which six hundred and ninety-five are cut and polished, many on both sides. The whole series, without exception, the author has had the honour of collecting for the noble donor.

The classification adopted is that of Dana's " System of Mineralogy," 1892, the pre-eminence being given to chemical composition. An exception to the arrangement involved in this rule is made in the case of the Fluorite group, which is placed *after* that of the oxides of Silicon, the author considering that, as the Collection is primarily one of Silica, they should, in this case, have priority of position. The Meteorites also, which would ordinarily take precedence under the " Native Elements," are placed at the end of the work before the Rocks.

The etymology of the name of each Mineral is given, together with its formula and quantitative analysis, the crystalline form, and the physical characters relating to cohesion, including hardness, specific gravity, cleavage, and fracture, with general reference to characters of diaphaneity, as well as the precise locality of each individual specimen. By referring to these data the student will be enabled to study and fully appreciate the more complex as well as the more simple characteristics of each respective species.

Free use has been made of Dana's Mineralogy (6th edition, 1892) as well as many other works, indicated by the reference notes.

The fact that this catalogue is also to serve the purpose of labels for the specimens in the Collection individually, has rendered repetition of the description in many cases unavoidable.

The author cannot conclude without acknowledging the kindness of the Chairman of the General Committee of the Liverpool Museum, Sir W. B. Forwood, as also of the Chairman of the Sub-Committee, W. H. Picton, Esq., who gave him every facility in arranging the collection at the Museum. To Mr. F. P. Marrat, who went over the nomenclature of the Devonian Corals individually ; as well as to Mr. Richard Paden, whose general assistance was of value, thanks are all due.   Mr. Chard's kind attention at the cases should also be acknowledged.

BRYCE WRIGHT.

London, December, 1893.

SILICA.

ROCK CRYSTAL, AMETHYST, ROSE QUARTZ, ETC.

SILICA is the dioxide of Silicon, and Quartz is its massive variety, which, when clear and transparent, is the *Crystal*, or Κρύσταλλος, of the ancients, and the rock crystal of modern days. Quartz is, without doubt, the commonest of all known minerals, and is distributed over the whole globe, either forming complete mountain ranges, such as the Sugar-loaf mountain of Wicklow in Ireland (which is of the whitest and purest description, occurring in a beautiful wave-like structure, refracting the light and looking like snow), or the complete range of mountains in the Sinaitic Peninsula, and other ranges too numerous to mention. It is also the primary and essential constituent of nearly all gravels with their pebbles ; the sands of the sea-shore,—all flints embedded in nodules or masses in the various strata of limestone (constituting the flint in the chalk formation), as well as the principal ingredient in metamorphic, as well as crystalline rocks, such as granite, etc. It also enters largely into the constituents of the earthy soils—argillaceous and calcareous—being of particular value in loosening these soils, without which they would become too stiff for agriculture

It is the principal ingredient also in mortar and cement, in the shape of sand, which is used largely for making glass, and

B

in other economic purposes, as well as in a state of combination with alumina, etc., constituting *Kaolin*—the principal ingredient of all *fine* porcelain. Not only is it the most essential of all minerals upon this earth, but probably forms a part
of the other worlds of " outer space." Professor Nevil Story-
Maskelyne discovered a form of dioxide of Silicon in the
interior of the Breitenbach Meteorite,* to which he gave the
name Asmanite, but which is now proved to be Tridymite,
a pure Silica (SiO_2).

The dimensions of Quartz in a pure state extend from a
line to hundreds of feet in thickness, as described by
Humboldt ; whole mountain ranges, as stated, being composed
of it, whose large rugged precipices under atmospheric in-
fluences become, in the course of time, disintegrated, primarily
crack, then are pressed open, until becoming detached, roll
down, carrying all before them, breaking and smashing into a
thousand angular fragments, which in their turn are worn
down by the ceaseless action of water until they form the
boulders, gravel-beds, and shingle of the rivers.

Quartz veins are often largely charged with the precious
metals,being nearly always either massive, or as a conglomerate, the matrix of gold and other valuable metals. The veins,
or lodes, in time become disintegrated by atmospheric agency,
pieces are worn away, revealing and laying bare the metals
within it ; doubtless discovered in this manner accidentally at
first, but afterwards systematically searched for. The massive
forms occur in great variety, and pass from the coarse to the
fine granular and crystalline kinds, to the crypto-crystalline or
flint-like masses, as well as occurring in mammillary, stalactitic,
and in concretionary and pulverulent forms, such as sand.

Man in his primitive state early recognised the value of the
flint-like masses, as illustrated by the many implements
discovered belonging to the Palæolithic epoch associated with
the remains of the extinct Mammoth, Woolly Rhinoceros, and

---

* " *Phil. Transactions*," 161, *page* 361, 1871.

other animals contemporaneous with man. These implements are invariably made of flint and quartzite, and nearly always of the same shape, viz., the most convenient for man to use with the hand—a solid round butt at one end, and pointed at the other ; whether discovered in the gravel pits of Norfolk or Suffolk, Abbeville, or in the Madras presidency, the shape is fundamentally the same. Passing from these weapons, which illustrate the first use to which any variety of Quartz was placed by man, we come to the crystallized variety known in modern times as Rock Crystal, and by the ancients as *Crystal.* The Greeks named it Κρύσταλλος—frozen or congealed, and evidently thought it was water permanently congealed through continued intense cold. Theophrastus, Aristotle, and particularly Pliny, were very positive of this, as the latter says, after alluding to the Murrhine vases :

" *Contraria huic causa* * *Crystallum facit gelu vehementius concreto*";

but *Diodorus Siculus*, if quoted correctly by Mr. G. W. Traill,† takes an opposite view, and thought crystal was water solidified by intense solar heat :

"*Crystallum esse lapidem ex aqua pura concretum, non tamen frigore sed divini caloris vi.*"

One of the early Greek writers, we are told, ascribed to Rock Crystal the power of producing the sacred fire of the Eleusinian mysteries, which was said to be kindled by placing it upon chips of wood and subjecting it to the sun's rays, which, causing the chips to smoke first, ignited them, producing a flame which was supposed to be grateful to the gods.

The Hebrews called it *Yahalom* from "halam," to strike, in allusion to Moses having caused water to spring out of the rock. The sixth precious stone in the breast-plate of the High Priest was a specimen of *Rock Crystal*, and was allotted to *Gad*, but under the name of *Adamas (Diamond)*, not, however, our well-known crystallized carbon, which was unknown

---

* " *Humorem putant sub terrâ calore densari.*"
† *Quartz and Opal,* G. W. Traill.

to them, but simply the quartz crystal, which they found to be very hard, not possessing the true Sapphire or Topaz at the time, which are almost the only gems, with the true Diamond that will cut or scratch it.

The large rounded masses of Quartz (see No. 1) are collected in the river beds of Minas Geraes, Brazil, Uruguay, Paraquay, and other South American localities and are called " Pebbles." It may be as well here to define that the term "pebble" is only used and restricted to water-worn specimens, produced by attrition on the beaches or in the streams, not to any silica, which has a globular or amygdaloidal form, derived from some cavity in which it may have been deposited, or owing to its own separate formation. A very wonderful specimen of Rock Crystal is in the collection (No. 12), exhibiting crystalline growth—the apex of a crystal in the interior showing different stages of formation clearly defined by a dusty deposit which is disseminated over each layer. The prism as well as the pyramid has been unfortunately polished, but permits its formation and growth to be more easily studied.

Rock Crystal is colourless, or nearly so, whether distinctly crystallized or not, although many of its varieties are of beautiful colours, such as the Amethyst, Rose Quartz, and others, the colouring matter being due to traces of metallic oxides or foreign earthy minerals. The pebbles found abundantly at Minas Geraes in Brazil and Madagascar are exceptionally clear and pellucid, and are extensively used for the manufacture of spectacles and eye-glasses, large quantities being imported into Europe annually.

All the varieties are employed more or less for ornamental purposes ; the pure, transparent, colourless specimens were largely used in the so-called " Cinque Cento " period,* being made into vases, tazze, bowls, etc., all most elaborately en-

---

* *The word Cinquecento, used so often by collectors, is the Italian abbreviation of " Mille cinque cento" (one thousand five hundred), the Sixteenth Century, during which period the Arts were greatly cultivated, after the commencement of the Medicean Age.*

graved. The ancients also used crystal for similar purposes, it being made into seals, cups, and chalices, and it was doubtless used for making many other ornaments. It is recited that the tyrant Nero dashed to pieces two magnificent rock crystal cups, one valued at £600, upon hearing of the outbreak of the insurrection which caused his downfall. During this Emperor's time, it was looked upon also as having medicinal properties.

Crystals of quartz are exceedingly beautiful, and are frequently found in groups; they occur also as lining the cavities of Geodes, or in incrustations, accompanying all the commoner metals. Enormous crystals are found not always white, but sometimes smoky. Specimens exist in the Berne Museum from Switzerland, perfectly crystallized, measuring over a yard in diameter. One from Brazil, in New York Museum, is 2ft. 6in. high, and over a foot in diameter, weighing 212lbs.; whilst another in the Museum at Paris is 3ft. in diameter, weighing 8cwt. A remarkable group of crystals is exhibited in the Museum at Naples, weighing half a ton, and another at Milan is $3\frac{1}{4}$ft. long, $5\frac{1}{2}$ft. in circumference, and weighs 870lbs. A Drusy cavity, about a century ago, was opened at Zinken, which contained no less than 50 tons of rock crystals (one weighing 800lbs.). The whole realized, even at that date, the sum of £60,000.

Large crystals are also to be seen in the British Museum, South Kensington, between the Mineral and Meteorite galleries.

The Green Mineral often seen in Rock Crystal is Chlorite, an admixture of silica, alumina, magnesia, and iron, with 10 per cent. of water, and is so plentiful at times that the Crystal is quite green, and may be mistaken by the uninitiated as being the true colour of the Rock Crystal caused by one of the metallic oxides, whereas it is only the intruding chlorite. Professor John Ruskin, in his "Catalogue of a Series of Specimens of Native Silica," suggests the name should be "Greenite" or "Greeny," in allusion to its colour; but the name Chlorite is very old, and is universally used, being found not only in Rock Crystal, but in granite, gneiss, diabase, and slaty rocks. It not only occurs green (in many shades), but there is, according to Phillips, a *blue* variety of Chlorite, so that the new suggested name can hardly be said to be *apropos.* Pliny knew the stone well, and described it as a stone of a grass-green colour—"*Chlorites herbacei coloris est* ' (Lib. xxxvii., 56).

Quartz crystals often contain crystals of foreign minerals, and are then termed SAGENITIC, from σαγήνη =- a net, the intruding crystals often crossing in different directions, forming reticulations like a net. Tourmaline, Epidote, Asbestos are some of the minerals, but the most beautiful and interesting of all is Rutile, the Dioxide of Titanium, which occurs in long golden or red acicular crystals, in bunches or isolated, running in every direction, in the purest pellucid quartz, a large and beautiful variety of which, through the indefatigability of the original collector, Mr. Francis Burton, are in this collection. They are called by the French "*Flèches d'amour*," "Love's Arrows," or "*Cheveux de Vénus*," "Venus' Hair." They are often found in the smoky variety of Quartz, termed "Cairngorm" (the *Mormorion* of Pliny*), who describes them thus :

"*Veneris crinis nigerrimi nitoris continet speciem rufi crinis.*"†

The beautiful bluish-violet coloured Rock Crystal called Amethyst has been known since the earliest Greek times. Pliny describes its colour very well as a purple gradually fading into light—

"*Ad viciniam crystalli descendit albicante purpuræ defectu.*"

It was called by the Hebrews ACHLAMAII and 'Αμέθυστος by *Theophrastus,* from α = not, and μεθύω = to intoxicate, because the ancients believed that wine drunk from an amethyst cup completely lost its power of intoxicating, as well as that it had the power of dissipating and preventing drunkenness. The Amethyst was assigned the ninth place in the breast-plate of the High Priest (1491 B.C.), allotted to

---

* *Pliny, Lib.* xxxvii., 63.　　　　† *Pliny, Lib.* xxxvii., 69.

the sixth and last son of Jacob by Leah, and had engraved upon it the name of the apostle "Matthias." It was the twelfth and last foundation-stone of the New Jerusalem, being symbolical then and since, in all times and countries of—humility.

It is often cut and faceted as a precious stone, exhibiting great beauty, particularly those gems which come from Siberia and Ceylon, generally being *dichroic*, exhibiting a beautiful pale or dark bluish-violet colour in the daytime, but turning to a deep garnet-red at night.

Its crystallization is naturally the same as Rock Crystal— rhombohedral, but there exists in the Amethyst fine strata, so superimposed that a triangular section is distinctly exhibited by cutting a thin slice of a crystal and examining it under the polarizing microscope, revealing as it is turned round the difference of colour caused by their arrangement and structure. Alternate layers exhibit rotary polarization right and left handed. A peculiarity in Amethyst are "feathers," which are very abundant, particularly in specimens from Brazil, which are caused, as is also the peculiar fracture in Amethyst termed "rippled," by these alternate layers. A very striking example in this collection is No. 54, which shows the hexagonal form of the crystal of a beautiful purple hue embedded in Amethyst, of a paler colour, illustrating crystalline growth. Other extraordinary examples are the Amethysts associated with Hornstone—Nos. 56 to 59 inclusive, lately discovered at Bergheim, Alsace.

The localities of the Amethyst are numerous, the finest most beautiful specimens coming from Brazil, Hungary, and Ceylon, and Siberia. Examples from Ceylon, as well as from Siberia, are sometimes erroneously called "Oriental" Amethysts, the true Oriental Amethyst not being a quartz at all, but a Sapphire—alumina. It occurs always in veins of the older formations, forming largely the interior of Geodes in the amygdaloidal trap rocks of Hungary, Silesia, Saxony, and the Tyrol, and many other parts of the globe. Many remarkable works of Glyptic art have been made in Amethyst. One of the most celebrated is a bust of Trajan in the Bibliothèque at Paris, whilst the Apollo Belvedere, the Farnese Hercules, and a small Laocoon Group in the same collection are exceptional works of art of the highest merit.

Its colour, which ranges from a pale amethystine tint, almost colourless, to a high and dark violet-blue, has always been attributed to oxide of manganese, but a German chemist, Heintz, has demonstrated that a very dark purple amethyst contained only 0·01 percentage of manganese, and that it loses its colour at 250°. In analysing another specimen he found it contained red oxide of iron, magnesia, lime, and soda. The presence of this latter alkali and the disappearance of the colour by heat, are in favour of Poggendorf's conjecture that the colour is due to the presence of a small percentage of ferric oxide. Whatever may be the cause, the beauty of an amethyst, either as a mineral or as a cut gem, is indisputable.

Another of the vitreous varieties of quartz is the CAIRN-GORM (often improperly called Topaz) or Yellow Rock Crystal, taking its name from the well-known Scotch locality Cairngorm in Banffshire, and used largely by the Scotch for jewelry as well as for the decoration of dirks.

The beautiful colour, by recent investigations by Forster, is said to be due to organic carbon-nitrogen compounds. Scotland has not yielded many specimens during late years, the stones sold in that country being principally from Ceylon and Brazil, where they occur in abundance. It was described by Pliny[*] as "*Mormorion*,' whilst the very dark nearly black specimens were called *Morion*.

Other vitreous varieties of Quartz are the "milky"— "*Quartz en chemise*" of the French—clouded crystals of a milky-white ; the cat's-eyes of Quartz (Crocidolite), as well as the interesting Aventurines, glistening with their scales

---

[*] *Lib.* xxxvii., 63.

of Mica, all of which are thoroughly represented in the " Derby Collection," and individually described.

Impure varieties of quartz from the presence of distinct minerals belong to the vitreous varieties, such as the *Eisen-kiesel*, Iron Flint, which, as its name indicates, consists of quartz crystals deeply impregnated with ferruginous sesqui-oxide (ferric oxide). Chloritic, actinolitic, and micaceous are other terms given to quartz containing various impurities derived from these respective minerals, whilst the term arenaceous is used to describe those of a sandy nature.

Rock crystals also often contain liquids which reveal their presence by small bubbles moving in a cavity which are dis-tinctly discernible on moving the specimen up and down. The water, or other liquid, is evidently surrounded and arrested within the crystals during their formation. Sir David Brewster made a series of experiments, and proved in many cases that the liquid was not water, but a fluid probably of an oleaginous nature, about twenty times thinner than water, which dis-appeared if the crystal was subjected to great heat, but reappearing on being cooled. One part of the liquid was volatile at twenty-seven degrees, but the other part was a fixed oil. Many rock crystals contain a naphtha fluid of a bituminous nature, but the most common crystals contain generally water with air bubbles, particularly those from Porretta, Bologna, Italy, of which several specimens are to be seen in the collection (Nos. 138 to 141 inclusive).

Passing from the pheno-crystalline or vitreous varieties of quartz, we arrive at the crypto-crystalline or flint-like massive varieties, which include the Chalcedonies, Agates, Jaspers, and other well-known varieties of silica which compose the bulk and most important section of the " Derby " Collection.

CRYPTO-CRYSTALLINE OR FLINT-LIKE MASSIVE VARIETIES.

CHALCEDONY, AGATE, JASPER, ETC.

THE varieties, Chalcedony, Agate, Jasper and other beautiful forms of native silica have been a favourite study from the earliest ages, either in their native state, or worked into the marvellous Camei and Intaglie left to us from the most remote periods—evidence existing to prove that as early as the Archaic Greek (three or four hundred years before Christ), engraving was executed by Phrygyllus in variegated stones; whilst the Assyrian cylinders and Egyptian scarabei, made of beautiful varieties of chalcedony and agate, prove that such stones were appreciated in those countries for their natural beauty. The taste spread eventually through Persia, Greece, and Etruria. The primary period of art in connection with stone may be said to have risen five hundred years before Christ and to have lasted for five hundred years afterwards, although long before that date colossal and gigantic works in stone, such as the Pyramids of Cheops, had been executed. The Onyx is first mentioned in the Bible* and also in the inscription of the Parthenon dating from the Peloponnessian War, B.C. 404 to 431. Theophrastus mentions the Agate (and this name included many varieties known to us now under various names), and that it derived its name from the river Achates (now called Drillo) in Sicily, always celebrated for its pebbles of silica. It has, however, been suggested that more probably the name was derived from Accho or Akka, the western section of the great plain of Esdraelon in Issachar, a country

* *Gen.* ii. 12.

described as " rich in minerals." The word Accho also signifies " heated sand," and Agates may be described in one sense as being made from siliceous sand fused into masses by igneous agency.

The Agate is mentioned in Scripture, 1491 B.C., as being the eighth stone in the breast-plate of the High Priest and was called *Issachar*, and it is generally admitted that this stone 'Αχάτες and Achates (Vulgate) is identical with our Agate. It was called Shebo in Hebrew, and was certainly known in its translucent if not transparent state, for the prophet Isaiah* says, " and I will make thy windows of agates." It was also used for other decorative purposes.

The formation, crystalline form, physical and other characters of this variety of silica is treated upon specially in a subsequent chapter preceding the individual description of the Agates, as is also the Onyx, Jasper, and other distinctive varieties.

The Onyx was spoken of by Job as the " precious onyx,"† a proof of the esteem in which it was held. This beautiful form of silica, of which there are abundant examples in this collection, is spoken of as being in the Garden of Eden,‡ and was the eleventh stone in the breast-plate of the Ephod and contained the name *Manasseh*, which signifies " That is forgotten " or " That makes forget." Its Hebrew name, *Shoham*, does not throw any light upon the interpretation in the word *Manasseh*, nor unfortunately does its Greek name, "Ονυξ, which signifies a nail (because the white portion of the stone resembles the nail growing out of the flesh) assist in any way. St. John mentions the Sard-onyx, which consists of reddish layers of Sard or Carnelian alternate with others, white or brown, as constituting one of the foundations of the Heavenly Jerusalem.§

Jasper, ' Iaspis,' or in Hebrew Jaspeh, another and most important silica, was the first foundation-stone of the New Jerusalem and bore the name of the Apostle Peter as well as

---

* *Isaiah* liv. 12.   † *Job* xxxviii. 16.   ‡ *Gen.* ii. 12.   § *Rev.* xxi. 20.

occupying the twelfth and last position in the breast-plate, with the name Ephraim and was considered of the greatest value, it being recited that during the existence of the Second Temple the Jaspeh of Benjamin was lost, and that to replace it Dama ben Nethinak was paid one hundred gold denarii, or about £60, for one somewhat similar. It is first mentioned in Exodus xxviii. 20, 1491 B.C.

The *Jasp-Achates* of the ancients, and mentioned by Pliny and other writers, is our Jasper-Agate ; the *Sard-Achates* of Pliny is our Agate containing layers of Sard or Carnelian, whilst the *Dendrachates* of Pliny, from δένδρον = a tree, correspond to our beautiful Moss Agates and Mocha Stones. *Hemachates*, from αἷμα = blood, which was an Agate sprinkled with spots of red jasper, is most probably the Bloodstone or Heliotrope. The Scriptures refer in many places, too numerous to mention, to varieties of Silica, proving the interest and veneration with which they were held at that epoch.

A most interesting form of silica are the agatized and siliceous woods, corals, and sponges, exhibiting clearly their formation and septa, although so changed from their natural state, for nearly all the silicified corals were primarily composed of carbonate of lime and have passed through the process of being changed into hard silica (in fact, the form they take is that of Hornstone, the hardest of all silicas). The species of the coral when distinct can easily be recognized and named, giving a proof that organic remains (for the sponges and corals and other zoöphytes are now by all advanced zoologists admitted into the animal kingdom) turn by process of time into silica, and the theory advanced that all flints may primarily have been of organic origin.

The West Indies seem to have undergone some wonderful and marvellous change during the growth of large forests of trees, particularly Antigua, where thousands of trunks, representing many species of silicified trees and palms are found, some of them being nearly two feet in diameter. Lately, Arizona, in America, has also produced some wonderful silicified woods, as well as many parts of India. The agate, or chalcedony, of different tints and colours intermingling and assuming the structure of the wood are very interesting and beautiful.

In a rocky part of this country, close to Chalcedony Park, a wonderful natural bridge exists, formed entirely of agatized wood—a centre of great attraction for tourists. The supposition is that at a remote period the tree had fallen and become embedded in the silt of an inland sea, that the silt in time became silicified, and that silex in solution poured into the pores, completely enveloped the tree, which passed gradually through all the stages of mineralisation until it became solid agate, exhibiting perfectly, however, its structure, pores, and bark in all the characteristic colours of true agate and jasper.

The interesting forms of silica which may be mentioned here are the *Mocha Stones* and *Moss Agates* or Dendrachates of Pliny. The Mocha Stones are specimens of translucent or semi-transparent varieties of white or bluish-white Quartz (Chalcedony), upon which black, brown, and red dendritical figures are displayed.

Being greatly sought after for making the tops of snuff-boxes and other objects, they are sometimes imitated by taking a piece of chalcedony of the required colour, and etching upon it by the aid of wax, honey, and sulphuric acid, or they are sometimes drawn upon with nitrate of silver or even marking-ink. The deception is easily, however, detected, as the tree-like figures are naturally always and only on the surface, as well as being faint, whilst the markings upon the true Mochas are seen to dip obliquely, and in many cases go right through the stone. They are called Mocha Stones from the first specimens having been found at Mocha, in Arabia. They were found principally in India, whilst cutting the section for the railway through the Vindyah mountains. Very large Mochas are unusual, one to three inches being the general size.

The specimen No. 381 from the collection of the late Arthur

Wells, Esq., of Nottingham, who procured it from Dr. Bird-wood, is one of the most clear and distinct known.

The Moss Agate is an exceptionally beautiful stone, com-posed of transparent chalcedony, in which is embedded beautiful moss, or lycopodium like dendritic markings of a most delicate description, in green of various shades with red, supposed for many years to be of a vegetable origin, but really arising from the precipitation of metallic oxides of manganese and iron. One of the most remarkable specimens known belongs to the Derby Collection, No. 366, exceptional for its great beauty by transmitted light and through the tree-like dissemination of the metallic oxides as well as for its abnormal size.

Heliotrope or Bloodstone, a very beautiful and interesting variety of silica, is a mixture of chalcedony with earthy chlorite and spots of ferric oxide, deriving its name from the Greek ἥλιος -- the sun, and τροπή a turning.

It is a well-known species on account of the red spots contained in it, and was highly valued in the Middle Ages by the superstitious, who regarded the red spots as Christ's blood diffusing through the stone. Many cameo-cutters have repre-sented Christ upon the Cross in heliotrope, arranging so that the red spots should only appear upon the wounds in the hands and side. One celebrated work of art of this description is in the Gallery of the Louvre.

The ordinary Bloodstone is opaque, and often when viewed with a magnifying glass will be seen to be of a granular or pisolitic texture, but a rare variety, associated with Jasper and classified under that name, not often met with, is quite trans-lucent, almost transparent, and is well represented in this collection by the beautiful specimens, Nos. 494 and 495.

Another green variety of silica which deserves special mention is the Chrysoprase. The ancients called a stone *Chrysoprasus*, but it is doubtful whether our stone is the same as the description given by Pliny does not agree with it.

The modern is generally of an apple, grass-green, or

whitish-green colour, caused by a small percentage of oxide of nickel. In the fourteenth century it was used for ornamenting churches and other places, it being then imported, most probably, from Ceylon and India. A great deposit of it was discovered in Silesia in 1740, by an officer, which locality even now furnishes specimens sought after greatly through its *refreshing* colour, which is of a beautiful and delicate light green.

Frederick the Second was so charmed with the mineral that he had his palace at San Souci ornamented with it.

Now its use is revived for jewelry purposes, although it is not worn so much as it was about a century ago, when oval specimens of from one half to two inches gradated were strung on gold chains and used as necklaces.

Other crypto-crystalline varieties of Quartz are the Touchstone, the *Basanites*, or *Lapis Lydius*, of Pliny, so well-known and used for testing the quality of gold with the aid of acid ; the Hornstone, and the well-known Flint or *Feuerstein*, so common and yet such a blessing to man in his savage days, as well as many varieties classified under the granular, conglomerate, and pseudomorphous varieties of Quartz.

The Precious, Fire, Wood, Rose, Prase and other Opals forming the last section of the Oxides of Silicon consist of specimens from Hungary, France, and other localities, but more particularly of examples of the Precious Opal from the comparatively newly discovered locality in Queensland—the Barcoo river—where they occur in an Iron-stone Jasper.

In Opal the Silica exists in a hydrated state, but in a different molecular condition to Quartz. The gravity is lower, as is also the hardness.

In the descriptive catalogue following, individual specimens are treated of in their order, whilst the formation of the beautiful and wonderful series of Agates, a collection quite unique in themselves, are treated of in a special chapter.

CLASSIFICATION.

OXIDES OF SILICON.

Quartz	SiO_2	Rhombohedral
Opal	$SiO_2 n H_2O$	Amorphous

MASSIVE QUARTZ—PURE.

No. 1. Pheno-crystalline, or Vitreous Varieties.

No. 2. Crypto-crystalline, or Flint-like, varieties.

A. PHENO-CRYSTALLINE, OR VITREOUS VARIETIES.

No. 1. *Ordinary Crystallized. Rock Crystal*, including (*a*) cavernous crystals, (*b*) drusy quartz, (*c*) globular quartz.

No. 2. Amethystine—Amethyst,'Αμέθυστον of *Theophrastus.*

No. 3. Rose quartz.

No. 4. Yellow quartz. False topaz or Citrine.

No. 5. Smoky quartz, Cairngorm stone, *Mormorion* of Pliny, 37.

No. 6. Milky quartz. "Quartz en Chemise" *Fr.*

No. 7. Sagenitic quartz, or quartz containing generally acicular crystals of (*a*) rutile, (*b*) göthite, (*c*) tourmaline, (*d*) asbestos, (*e*) chlorite, (*f*) actinolite, (*g*) hornblende, (*h*) epidote, (*i*) carbon.

No. 8. Cat's-eye. *Œuil de Chat. Fr. Katzen-Augen. Germ.* Including the Crocidolite (Faserkiesel).

No. 9. Aventurine.

No. 10. Impure varieties of Distinct Minerals, distributed entirely through the quartz, (a) *ferruginous, Eisenkiesel, Germ,* (b) *chloritic*, (c) *actinolitic*, (d) *micaceous.*

No. 11. Containing liquids in cavities.

B. Crypto-crystalline Varieties.

No. 1. Chalcedony. Murrhina, *Pliny*, 37, 7. Chalcedon *Germ.* Chalcedoine *Fr.*

No. 2. Sard or Carnelian. Sarda of *Pliny*, Σάρδιον of *Theophrastus*, Cornaline *Fr.*

No. 3. Chrysoprase, but not the antique Chrysoprasus.

No. 4. Prase from πράσον, a leek.

No. 5. Plasma. Iaspis, *pt. Pliny*, including Heliotrope, or Bloodstone.

No. 6. Agate, ᾿Αχάτης of *Theophrastus* from the river in Sicily. (*a*) Banded, (*b*) irregularly clouded, (*c*) colour due to visible impurities and agatized wood—wood petrified with clouded agate.

No. 7. Onyx. ᾿Ονύχιον of *Theophrastus*. Pliny, 37, 24.

No. 8. Sardonyx, *Pliny*, 37, 23.

No. 9. Agate Jasper.

No. 10. Siliceous Sinter.

No. 11. Flint. Silex pt. *Pliny.*—Feüerstein *Germ.*

No. 12. Hornstone. Hornstein *Germ.*

No. 13. Basanite : Lydian or Touchstone. Lapis Lydius, *Pliny*, 33 and 43. Lydite.

No. 14. Jasper (*a*) Red Hæmatit*i*s, not his Hæmatit*e*s, (*b*) brown or ochre yellow, (*c*) dark green and brownish-green, (*d*) greyish-blue, (*e*) brownish-black or black, (*f*) riband or striped jasper, (*g*) Egyptian jasper, (*h*) jasponyx, (*i*) jasperized wood.

C. Quartz Rock or Quartzyte.

No. 1. Quartz granular.

No. 2. Quartz conglomerate.

No. 3. Pseudomorphous quartz. (*a*) Haytorite, (*b*) Babel quartz, (*c*) silicified corals, (*d*) pegmatyte.

Opal.

Opal. Pæderos, *Pliny*, 37, 21, 22

No. 1. Precious opal.

No. 2. Fire opal.

No. 3. Common opal. (*a*) Wax opal.

No. 4. Rose opal.   Quincite.

No. 5. Prase opal.

No. 6. Wood opal.   Lithoxyle.

No. 7. Fiorite or pearl sinter.

No. 8. Geyserite.

No. 9. Floatstone, Quartz nectique.

FLUORIDES.
Fluorite Group.

This group, according to the classification of the last edition of Dana, 1892, would really come before the Silicas, but it was thought that as they (the Silicas) constitute the greater part of the collection, and are of the most importance, that they should be placed first.

No. 1. Fluorite or Fluor Spar, Fluate of Lime, Fluoride of Calcium, "Blue John" of Derbyshire miners.

CARBONATES.
Calcite Group.

No. 1. *Variety based upon Crystallization.* (a) Calcite— Iceland spar.

No. 2. *Fibrous and Lamellar variety.* (a) Satin spar.

No. 3. *Granular Massive to Crypto-Crystalline varieties.* (a) Mexican onyx, (b) Calc Tufa, (c) Lumachelle marble, (d) Laminated oolite, (e) Ruin marble, (f) Limestone in which all Devonshire marbles and Devonian fossil corals are included.

Basic Carbonates.

Malachite.   Χρυσόκολλα *pt.* Theophrastus, carbonate of copper.

Azurite.—Lapis armenius of *Pliny*, 33, 57. Blue carbonate of copper associated with malachite.

SILICATES.

A. ANHYDROUS.

Feldspar Group.

Orthoclase. (*a*) Perthite.
No. 1. Microcline. No. 2. Amazonite. No. 3. Labradorite.

Pyroxene Group.

No. 1. Enstatite, Bronzite.
No. 2. Hypersthene, Labrador.
No. 3. Jadeite of Damour, Jade part.
No. 4. Rhodonite.

Amphibole Group.

No. 1. Nephrite from νεφρός—a kidney.
No. 2. Asbestos.

Beryl Group.

No. 1. Beryl, Aquamarine. (*a*) Emerald.

Iolite Group.

No. 1. Iolite or Dichroite, Cordierite, Peliom.

Sodalite Group.

ORTHOSILICATES.

No. 1. Sodalite.
No. 2. Lazurite or Lapis-Lazuli.

Garnet Group.

No. 1. Garnet.   (*a*) Carbuncle (Carbunculus).

Topaz Group.

No. 1. Topaz, not the Topazos of Pliny.

Epidote Group.

(Zoisite.)
(*a*) Thulite, a rose-red variety of Zoisite.

Prehnite.

An Orthosilicate not included in the Epidote group.
No. 1. Prehnite.

B. Hydrous Silicates.

Mica Group.

No. 1. Agalmatolite, or Pagodite.
No. 2. Lepidolite.   Lithia Mica.

Serpentine Group.

No. 1. Serpentine.   Massive (*a*) Bowenite.   Lamellar *b)*
Williamsite.   Fibrous (*c*) Chrysotile.

SULPHATES.

Barite Group.

No. 1. Barite—Sulphate of Baryta.   No. 2. Gypsum.

HYDROCARBON COMPOUNDS.

Oxygenated Hydrocarbons.

No. 1. Amber, or Succinite. Lyncurium.

NATIVE ELEMENTS.

Meteorites.

(*a*) Meteoric Irons, Aerosiderites or Siderites, (*b*) Meteoric Iron and Stone combined—Siderolites or Lithosiderites, (*c*) Meteoric Stones—Aerolites or Litholites.

Slickenside, an ungrouped metal, is placed in here.

ROCKS.

No. 1. Euphotide. No. 2. Porphyrite. No. 3. Antique green Porphyry. No. 4. Egyptian Diorite. No. 5. Variolyte. No. 6. Obsidian. No. 7. Granite.

OXIDES OF SILICON.

Quartz SiO_2. Rhombohedral, trapezohedral.
Opal $SiO_2 n H_2O$.

QUARTZ, the Κρύσταλλος of Theophrastus, etc. The crystallus
of Pliny, 37, 9, 10, who alludes to the hexagonal form with
pyramidal terminations, also mentioned by him as Silex
36, 371. Quertze, Kiselstein of Agricola, 276, etc. Cristal
de Roche *Fr.;* Berg-Krystall *Germ.;* Quarzo *Ital.;* Cuarzo
Span. The word Quartz is of provincial German origin.

Rhombohedral with trapezohedral tetartohedrism. Crystals
generally prismatic, often doubly terminated and distorted,
sometimes twisted or bent. Occurs lining the cavities of
Geodes, in druses or in radiated masses with a surface of
pyramids. Single crystals, Rose has shown, are right and
left-handed. Often in twin-crystals. The composition of
Quartz from an analysis made by Bucholz of a transparent
colourless crystal is Silica 99·37 with traces of Alumina.
Cleavage difficult, but may be obtained by plunging into
cold water after being heated. Fracture in crystallised forms
—conchoidal to sub-conchoidal—in some massive varieties
splintery and uneven. Brittle to tough. Hardness = 7·0 ;
cannot be scratched with the knife. Specific gravity, according
to *Beudant*, in crystals, 2·653, 2·654. The Herkimer county
crystals gave Mr. S. L. Penfield, of New Haven, a mean of
2·660. The flint-like massive varieties are lower—2·60 when
pure—but massive impure varieties, such as Jasper, are higher
according to the nature of the impurities. The extremes may
be given as 2·50-2·80. Lustre vitreous and sometimes greasy.
Splendent to nearly dull. When pure, colourless. but often

various shades of yellow, brown, red, black, violet, green, and blue. Streak white of pure varieties, and of impure a pale variety of the colour. Insoluble in all acids but hydrofluoric; infusible before the blow-pipe. Transparent and translucent to opaque.

Optical properties—Positive. Weak double refraction. Circular polarization ; hence the axial figure has a coloured centre. Rotation right and left-handed according to character of crystals. Pyroelectric ; also electric by pressure. Two pieces of Quartz rubbed together in the dark produce a phosphorescent light with a slight empyreumatic odour.

Composition Silica : or Silicon dioxide $SiO_2 =$

Oxygen	... 53·3
Silicon	46·7
	100·0

PURE MASSIVE QUARTZ.

No. 1. QUARTZ " PEBBLE," rough, water-worn specimen, caused by "attrition" and rolling in a river-bed stream, exhibiting internal fractures as well as on the surface conchoidal fracture. Transparent and colourless, but here and there tinged red in the interior. Polished on one side. Most probably detached from a mass of pellucid Quartz, primarily carried down to a river-bed and worn by rolling into its present shape. 5 by 3 ins.

River Bed of Minas Geraes, Brazil.

A characteristic specimen of the " Brazilian Pebble" so extensively used in the manufacture of spectacle and other glasses. The term "Pebble" is restricted to water-worn specimens of Quartz, and does not apply to nodules of any description whose shape may be owing to the cavities which they have filled in rocks.

No. 2. QUARTZ, massive, colourless to white, transparent, exhibiting prismatic colours—termed iridescence, caused by internal fracture and the interference of light in minute fissures, not in any way by the structure of the stone. Fracture conchoidal and uneven on the exterior. Cut and polished on one surface. 3 by 2½ ins.

Matumbaugh Mts., Madagascar.

No. 3. QUARTZ of a milky-white, semi-crystallized, exhibiting faintly the hexagonal forms of the crystals, associated with Chalcedony. Polished on one side. 3½ by 1¾ ins.

Galgen-Berg, Oberstein on the Nahe.

No. 4. QUARTZ, massive, white in feathery clouds, cut as an oblong paper-weight. Bevelled edges. Well polished all over. Transparent. 4⅛ by 3⅛ ins.

Matumbaugh Mts., Madagascar.

No. 5. QUARTZ, massive, white, clouded and feathery in the interior. Cut as a circular paper-weight. Polished all over. Diameter, 3¾ ins.

Matumbaugh Mts., Madagascar.

No. 6. QUARTZ, of a smoky tint, massive generally, but possessing crystallized planes with parallel striations on the prism. Exhibits most beautiful iridescence. Uneven and conchoidal fracture on the exterior. 7½ by 5 ins.

Matumbaugh Mts., Madagascar.

The name "Iris," signifying Rainbow, was given to specimens of Quartz cut out of pellucid crystals, exhibiting iridescence, and is mentioned by Pliny.* The iridescence or "iris" is the effect of colour similar to that produced by the presence of a small amount of air between two thin plates pressed together, which becomes expanded more or less by pressure, and which in the Quartz is arrested between interior crevices. "Rubasse" is another name given to specimens of pure Quartz plunged, after being repeatedly heated, into a solution of cochineal, which enters into the fissures. Natural "Rubasse" is said to occur at Minas Geraes in Brazil, but the author has never seen it in Europe, although the artificial specimens are common in several tints.

* *Lib.* xxxvii., 52.

A. Pheno-crystalline or Vitreous Varieties.

ORDINARY CRYSTALLIZED ROCK CRYSTAL.

No. 1. *Rock Crystal.*

No. 7. Group of Rock Crystals, slightly smoked, transparent. The crystallization of the pyramids is very interesting, one plane being largely developed at the expense of the others. Composition SiO_2. $9\frac{1}{2}$ by 6 ins.

Dauphiné, France.

Coll. Henri de Laurençel of Fontainebleau.

No. 8. Group of Rock Crystals of a slight reddish tinge. Translucent. Hexagonal form with pyramids well exhibited with drusy base. $6\frac{1}{2}$ by $3\frac{3}{4}$ ins.

Caveradi, Tavetschthal, Switzerland.

No. 9. Group of Rock Crystals, very brilliant, transparent. Many of the pyramidal planes distorted, associated with Albite, one of the Feldspars. 5 by 4 ins.

Tintagel mine, Cornwall.

No. 10. Group of Rock Crystals, pure white on a massive Quartz and drusy base. Very beautiful specimen. 6 by 4 ins.

Coll. Arthur Wells. Snowdon, Wales.

No. 11. Group of Rock Crystals, transparent. Some of the planes of the pyramid developed at the expense of others, like the Dauphiné specimens. Parallel striations upon the prism. Transparent in part, but feathery. $5\frac{1}{4}$ by $2\frac{1}{2}$ in.

Maderanerthal, Uri, Switzerland.

No. 12. Rock Crystal, pure pellucid specimen exhibiting in the interior stages of growth of another crystal the layers of the new planes of the pyramid of which are discernible partly through the dissemination of a "fine dust" upon each stratum.

The crystal has been unfortunately cut a little out of its natural shape, and polished all over. $2\frac{3}{4}$ by $1\frac{3}{8}$ ins.

Coll. David Forbes. Minas Geraes, Brazil.

A wonderful specimen illustrative of internal crystalline growth.

No. 13. QUARTZ CRYSTAL, pure pellucid with one plane of the pyramid developed at the expense of the others; the prism exhibits parallel striations. Transparent. 3⅜ by ¾ ins.

Dauphiné, France.

No. 14. Group of ROCK CRYSTALS, exhibiting manner of attachment, striated with large developed pyramid. 3 by 2 ins.

Tavetschthal, Switzerland.

No. 15. Group of ROCK CRYSTALS, very brilliant, with massive quartz base, associated with Adularia Feldspar. Transparent. 3½ by 2¼ ins.

Tintagel mine, Cornwall.

No. 16. Group of QUARTZ CRYSTALS, pure white feathery on a base of Chalcedony with Chlorite. 2¼ by 2 ins.

Oberstein on the Nahe, Germany.

No. 17. ROCK CRYSTALS, pure transparent, embedded in a cavity of Carrara marble, with drusy quartz. The fine white statuary marble used by modern sculptors. 4¼ by 2 ins.

Carrara, Italy.

Coll. Marquis de Chigi.

No. 18. Pellucid crystal of QUARTZ, colourless, with a little Chlorite on a drusy built-up base. 2¾ by 1 in.

Tavetschthal, Switzerland.

No. 19. Group of ROCK CRYSTALS, very brilliant, with distorted prisms, white, transparent. Tinged with iron oxide at base. 2½ by 1¾ ins.

Tintagel mine, Cornwall.

No. 20. QUARTZ CRYSTALS, pure white, upon a base of pale green Fluor Spar (Fluoride of Calcium). 5¼ by 4¼ ins.

From the Alston Moor district, Cumberland.

No. 21. QUARTZ CRYSTALS formed upon cubes of Fluor Spar. Milky-white, very brilliant. Fluor Spar showing at side. 4¾ by 3½ in.

District of Weardale, Cumberland.

No. 22. QUARTZ CRYSTALS, white and brilliant, upon cube of blue Fluor Spar, with Galena. 4 by 3 ins.

District of Weardale, Cumberland.

No. 23. QUARTZ CRYSTALS, pure white, associated with Fluor Spar (Fluoride of Calcium). 4½ by 2½ ins.

District of Weardale, Cumberland.

No. 24. QUARTZ CRYSTALS, pure white and brilliant, associated with Calcite (Carbonate of Lime), in lenticular crystals. 4½ by 3¼ ins.

Alston Moor district, Cumberland.

No. 25. QUARTZ CRYSTALS, pure white with lilac-coloured Fluor Spar (Ca. Fl.). 3¼ by 2¾ ins.

Alston, Cumberland.

No. 26. QUARTZ, crystallized, pure white and very brilliant, upon amber Fluor Spar. 3 by 2¾ ins.

Alston Moor, Cumberland.

No. 27. QUARTZ, crystallized, milky-white, associated with crystals of Calcite (Carbonate of Lime). 2⅛ by 2 ins.

District of Alston, Cumberland.

No. 28. QUARTZ, crystallized, exhibiting an interesting "division," or "cut," sometimes met with in specimens from the iron mines, caused through thin laminar crystals of specular iron which have become disintegrated and fallen out, associated with Pearl Spar—a variety of Dolomite—upon Hæmatite, the iron ore of commerce. 3 by 1¾ ins.

Cleator Moor, Cumberland.

No. 29. QUARTZ, crystallized, brilliant aggregation of crystals associated with Hæmatite. 2½ by 2¼ ins.

Iron mines of Cleator Moor, Cumberland.

No. 30. QUARTZ CRYSTAL (single), slightly smoked upon a base of Hæmatite. 2½ by 2¼ ins.

Cleator Moor, Cumberland.

No. 31. ROCK CRYSTALS, pure pellucid, doubly terminated, very perfect. Transparent, enclosing black grains, probably of Carbonaceous matter. The largest is 1¼ by ⅞in.

Fairfield, Herkimer co., N.Y., U.S.A.

No. 32. QUARTZ CRYSTALS. Eight specimens, pure transparent, some doubly terminated, detached from a cavity in Carrara marble. Small, in glass tube.

Carrara, Italy.

No. 33. QUARTZ. Two crystals joined together. Doubly terminated. Exhibiting cavities caused by the impression or intrusion of other substances, in this case probably clay, which has became disintegrated. 3¼ by 2½ in.

Porretta, Bologna, Italy.

No. 34. QUARTZ CRYSTALS. A group exhibiting an aggregation of crystals, some doubly terminated, others distorted with parallel striations on the prisms. White. Transparent to translucent. Originally embedded probably in clay, portion of which is attached to the crystals. 3½ by 3⅛ ins.

Porretta, Bologna, Italy.

Coll. Prof. Louis Bombicci.

No. 35. CRYSTAL of QUARTZ. White. Doubly terminated, distorted. Two crystals are developed at one extremity. Translucent but feathery. 2½ by 15⅛ ins.

Porretta, Italy.

Coll. Prof. Louis Bombicci.

No. 36. ROCK CRYSTALS, isolated, three. Doubly terminated with the planes greatly distorted. White. Pellucid, but feathery. Prisms exhibit parallel striations. 1 to 1½ ins. long.

Porretta, Italy.

Coll. Prof. Louis Bombicci.

No. 37. QUARTZ CRYSTALS. An important group. Cavernous, caused probably through the impression of other minerals which have become disintegrated. Exhibiting remarkable growth of crystals, forming by aggregation an irregular doubly terminated specimen.

The peculiar structure is termed " en trémies " or " hopper" like, "hopper" being a mining term applied to specimens which exhibit a cubical step-like formation.

Translucent, with a glassy fractured appearance, looking *semi-hacked.* 3 by 3 ins.

Porretta, Italy.

The description " en trémies" is used generally in Italy for this variety of Quartz, but the term " hopper" is used in other parts of Europe and England to describe a series of steps which may often be seen produced by evaporation in some of the Haloids.

No. 38. QUARTZ CRYSTALS. Aggregated together. The prisms and pyramids of this specimen are cavernous to a marked degree, to which the term " en trémies typique " or " hopper-like" has been given by Professor Louis Bombicci of Bologna.

The cavities enclose clay, and the specimen represents well its hydrothermal origin. Glassy and in parts transparent. 2¾ by 2¼ ins.

Du grés de Porretta, Italy.

Coll. Prof. L. Bombicci.

No. 39. ROCK CRYSTAL, built up, exhibiting parallel striations in the prism, with clay embedded in the pyramid. Exhibits a movable drop of water with air-bubble. Distorted. Transparent. 2¼ by ⅝ in.

Porretta, Bologna, Italy.

No. 40. Group of ROCK CRYSTALS, "en trémies," enclosing clay and exhibiting growth of crystals. Fairly white. Greasy look. Transparent to translucent. 3 by 2¾ ins.

Porretta, Bologna, Italy.
Coll. Prof. L. Bombicci.

No. 41. Group of ROCK CRYSTALS, "en trémies," colourless to white, distorted, built-up crystals. White. Glassy—transparent to translucent, exhibiting iridescence and hydrothermal origin. 3¼ by 2 ins.

Porretta, Italy.
Coll. Prof. L. Bombicci.

No. 42. Group of ROCK CRYSTALS, white, " en trémies," or hopper-like. Enclosing clay and exhibiting hydrothermal origin. Semi-hacked through disintegration probably of a foreign substance. Glassy in appearance. Translucent. 3¼ by 1½ ins.

Porretta, Italy.
Coll. Prof. L. Bombicci.

No. 43. QUARTZ CRYSTALS, " en trémies " or "hopper " like enclosing clay of a light brown colour and exhibiting hydrothermal origin. White generally, glassy in appearance and semi-hacked. 2½ by 2 ins.

Porretta, Bolonais, Italy.

No. 44. Aggregation of ROCK CRYSTALS " en trémies." enclosing clay. Hacked and striated. 2½ by 2¼ ins.

Porretta, Italy.

No. 45. Two small ROCK CRYSTALS, "en trémies," planes hacked, prisms exhibiting parallel striations. 1½ and 1 in.

Porretta, Italy.
Coll. Prof. Louis Bombicci.

No. 46. ROCK CRYSTAL, transparent, planes of the pyramid irregularly developed, sub-conchoidal fracture shown at base. Exhibiting parallel striæ. 2 by ¾ ins.

Traversella, Piedmont.

No. 47. QUARTZ CRYSTALS, drusy, so called by being incrusted with minute crystals, associated with Fluor Spar (Fluoride of Calcium). 5 by $2\frac{1}{2}$ in.

Alston Moor, Cumberland.

No. 48. DRUSY QUARTZ, small crystals disseminated over pyramids, larger crystals on massive Quartz base. Pure white and extremely brilliant. $2\frac{1}{2}$ by $2\frac{1}{4}$ ins.

Alston, Cumberland.

No. 49. QUARTZ CRYSTALS, aggregation in dodecahedrons, with drusy faces, "*faces sagrines.*" $1\frac{3}{4}$ by $1\frac{1}{4}$ ins.

Montecuto, Bologna, Italy.

No. 50. QUARTZ, radiated, of a greenish tinge. Found in the ophisilicates and serpentines. 3 by 2 ins.

Lizzo, Bolonais, Italy.

Coll. Prof. Louis Bombicci.

No. 51. QUARTZ in sphærohedrons, pure white, upon concretionary limestone, bituminiferous with native sulphur. $3\frac{1}{2}$ by $2\frac{1}{4}$ ins.

Sulphur district of Cesena Romagna, Italy.

No. 52. QUARTZ, crystallized globular, a rare variety, associated with native crystallized sulphur upon limestone. $2\frac{3}{4}$ by $1\frac{3}{4}$ ins.

Sulphur mine of Romagna, Italy.

No. 53. CRYSTAL QUARTZ, with a little Chlorite. Doubly terminated. At one extremity is a rather broad flattened crystal of Rutile. $1\frac{3}{4}$ by 1 in.

Canton des Grisons, Switzerland.

AMETHYST QUARTZ.

No. 54. AMETHYST QUARTZ. Section of a beautiful and brilliant clear purple or bluish-violet. Translucent and very lovely by transmitted light. Illustrates in a marked degree crystalline growth—the hexagonal form of a deep violet colour surrounded by a lighter hue paling off towards the extremities. Feathery. Cut and polished on both sides. The colour is said to be due to Ferric Oxide (*see* page 8). An analysis of Amethyst Quartz by Rose yielded :

Silica...	97·50
Alumina	0·25
Oxide of Iron ...	0·50
Oxide of Manganese	0·25
	98·50

6 by 5 ins. and ½ in. thick.

Minas Geraes, Brazil.

A remarkably beautiful specimen, both for colour and crystallization. It exhibits also the peculiar "rippled" fracture caused by minute alternate layers, as well as "feathers," which can be plainly discerned. One of the gems of the collection.

No. 55. AMETHYST QUARTZ. Thick section of a very beautiful deep violet-blue or clear purple of two distinct tints, cut across the centres of closely aggregated crystals, some of which clearly exhibit their hexagonal form. Polished on one side. 5½ by 3 ins. and about 1½ in. thick.

Minas Geraes, Brazil.

No. 56. AMETHYST QUARTZ. Large and thin section cut from a massive block, associated with other minerals. Centre, an opaque-brown Jasper, surrounded by bands of Amethyst and crypto-crystalline Quartz, encircled by strata of white Jasper and Agate in small fortifications, with amorphous Silica and red Jasper on the exterior. Partly opaque, but translucent also in some parts by transmitted light. Polished on both sides. 9½ by 8 ins.

Bergheim, Alsace, Germany.

A comparatively new but beautiful variety of Amethystine Quartz, discovered in large masses a few years ago in Alsace, Germany. It has been used successfully for making table-tops, pedestals, tazze, cups, and other ornaments, a large series of which are in the possession of the distinguished collector, Mr. Alfred Morrison.

No. 57. AMETHYST QUARTZ. Another section similar to No. 56. The hexagonal form of the prisms is seen by transmitted light. A paler specimen than No. 56. 8¾ by 6½ ins.

Bergheim, Alsace.

No. 58. AMETHYST QUARTZ. Section of a beautiful violet-blue colour at one side, with crypto-crystalline silica, opaline by transmitted light, associated with Agate in small fortifications. Polished on both sides. 7 by 6 ins.

Bergheim, Alsace.

No. 59. AMETHYST QUARTZ, exhibiting hexagonal form of crystal by transmitted light ; a fresh violet colour. Associated with white Jasper and Agate. 5 by 4 ins.

Bergheim, Alsace.

No. 60. AMETHYST QUARTZ SECTION. The centre of semi-crystallised Amethyst of a fresh violet colour, surrounded by concentric layers of white and grey-blue Chalcedony. From the amethyst centre there is an " inlet of infiltration " or an " outlet of egress." Polished on one side. 4¼ by 3 ins.

Minas Geraes, Brazil.

No. 61. AMETHYST QUARTZ. Small specimen, translucent, cut and polished on both sides. 2 by 1¼ ins.

Minas Geraes, Brazil.

No. 62. AMETHYST QUARTZ. Group of crystals of a pale violet tinge. The prisms and pyramids well defined. 3 by 2¼ ins.

Schemnitz, Hungary.

No. 62A. AMETHYST QUARTZ, cut into the shape of an egg, of a beautiful amethystine colour, associated with bands of crypto-crystalline Silica, Agate, and Chalcedony. Length 2½ ins. Cut and polished all over at Oberstein.

Minas Geraes, Brazil.

ROSE QUARTZ.

ROSE QUARTZ. A peculiar but striking variety of Silica, so called from its colour. It is always massive, and presents a peculiar cracked appearance on the surface, and occurs associated with the rocks Granite and Gneiss. Its colour was said to be due to Oxide of Manganese, but a German mineralogist, Prof. Fuchs, has lately demonstrated that Rose Quartz by getting pale, as it always does if exposed to the light, would not contain Manganese; and that the specimens he has analysed from Rabenstein—its principal locality—owe their colour to the presence of 1·0 to 1·5 per cent. of Oxide of Titanium, a mineral, under the name Rutile, very often found in acicular crystals impregnating Quartz, and well represented in this collection. Its unusual colour causes this massive Quartz to be eagerly sought after for making into works of art. Busts, vases, and other objects are frequently of special value if the colour be fine. A magnificent vase, no less than nine inches high, formerly in the possession of the Marquis de Drèe, is quite unique.

———

No. 3. Rose Quartz.

No. 63. ROSE or PINK QUARTZ. Massive specimen, polished on one side ; much cracked on the surface. The colour is said to be due to Manganese, but Fuchs attributes it to 1 to 1½ per cent. of Titanium. Well polished. 4½ by 3½ ins.

Rabenstein, Bodenmais.

No. 64. ROSE QUARTZ, light pink, massive, surface cracked, which is a peculiar characteristic of this mineral. Similar specimen to No. 63. 4½ by 3½ ins.

Rabenstein, Bodenmais.

No. 65. ROSE-RED QUARTZ of a fine deep pink colour on one side   Cut and polished.   3½ by 3¼ ins.

Tavastchus, Finland.

No. 66. ROSE QUARTZ, pale pink colour.   Cut and polished. 2¼ by 1¼ ins.

Rabenstein, Bodenmais.

No. 4.   Yellow Quartz, False Topaz, or Citrine.

No. 67. CITRINE, FALSE TOPAZ, or YELLOW QUARTZ, showing iridescence and conchoidal fracture.   2¼ by 1¼ ins.

Matumbaugh Mts., Madagascar.

No. 5.   Smoky Quartz.

No. 68. SMOKY QUARTZ CRYSTAL. *Rauch Quarz*, Germ. Deep black, opaque.   The *Morion* of Pliny—a name given by him to this very dark or black coloured Quartz. *Mormorion* is the name given to a lighter or Cairngorm colour.   A long crystal with pyramid attached to another, the prism horizontally striated.   7½ by 2 ins.

Ruäras, Sidnun-Tavetschthal, Switzerland.

No. 69. SMOKY ROCK CRYSTAL, CAIRNGORM, *Mormorion* of Pliny.   Transparent, feathery, two crystals attached 2¾ by 1½ ins.

Tavetschthal, Switzerland.

No. 70. QUARTZ CRYSTALS, slightly smoked upon mammillated Hæmatite, the iron ore of commerce, associated with Pearl Spar.   5 by 3 ins.

Cleator Moor, Cumberland.

No. 71. QUARTZ, smoky crystals, some tinged red with Oxide of Iron upon Hæmatite.   4¼ by 3 ins.

Cleator Moor, Cumberland.

No. 72. QUARTZ CRYSTALS, slightly smoked, associated with micaceous iron. 3¼ by 2¼ ins.

Cleator Moor, Cumberland.

No. 73. SMOKY QUARTZ, with "*hacked*" quartz, so called from its peculiar "hacked" appearance, the cavities causing this structure being formerly occupied by a substance which has disappeared through disintegration. In this specimen the cavities at the back are covered with minute brilliant drusy crystals, associated with micaceous specular iron. 4½ by 2 ins.

Cleator Moor, Cumberland.

No. 74. SMOKY or SMOKED QUARTZ. *Rauch Quarz.* With the pyramidal faces well developed, associated with micaceous specular Iron and Hæmatite, the iron ore of commerce. Partly botryoidal. 3 by 3½ ins.

Cleator Moor District, Cumberland.

No. 6. Milky Quartz.

No. 75. MILKY QUARTZ. "*Quartz en chemise*," *French.* Crystallized on massive white quartz. Greasy lustre, sometimes termed *Greasy Quartz.* 3¼ by 2¼ ins.

Snowdon, Carnarvonshire.

No. 7. Sagenitic Quartz.

SAGENITIC QUARTZ (from Σαγήνη, a net), containing within it thin, slender Rutile (L., *Rutilus* = shining red) crystals which, crossing each other, become reticulated or net-like. The following foreign substances also found in Quartz are under the same classification: (*a*) Rutile, (*b*) göthite, (*c*) tourmaline, (*d*) asbestos, (*e*) chlorite, (*f*) hornblende, (*g*) epidote, (*h*) carbon.

No. 76. QUARTZ, white, enclosing isolated as well as bunches of parallel acicular crystals of resplendent golden Rutile (the Oxide of Titanium), penetrating the Quartz in every direction. The Sagenite of Saussure. The *Chrysothrix*, or golden hair of the Orphic Poem. Called by the French "*Flèches d'amour*" or "Love's darts." Also "Venus' Hair," in allusion to its resemblance to golden hair. Polished all over. 4¼ by 4 ins.

San Juan del Rey, Minas Geraes, Brazil.

A wonderful and probably unique specimen, both for beauty and the mass of needle-like golden crystals it contains.

No. 77. QUARTZ, colourless, pure pellucid, containing acicular crystals of Rutile (TiO_2). Exhibits internal fracture. Variety called "Brazilian Pebble." Polished all over. 4½ by 2½ ins.

San Juan del Rey, Minas Geraes, Brazil.

No. 78. QUARTZ, perfectly pure and transparent, showing internal fracture with isolated acicular crystals of Rutile. " Venus' Hair," or "*Flèches d'amour*" (" Love's darts "). Cut and polished all over. 4 by 3¼ ins.

San Juan del Rey, Minas Geraes, Brazil.

A specimen remarkable for its limpidity.

No. 79. QUARTZ, pure pellucid specimen with needle-like crystals of Rutile. *Chrysothrix* of the Orphic Poem. Exhibiting internal fractures. Cut and well polished all over. 5 by 2 ins.

San Juan del Rey, Minas Geraes, Brazil.

No. 80. QUARTZ, quite pure, containing bunches of Rutile, the Oxide of Titanium. *Chrysothrix* of the Orphic Poem. A conical shape. The acicular crystals, which run parallel, are gathered at the top. Polished all over. 3½ by 2¾ ins.

Minas Geraes, Brazil.

No. 81. QUARTZ, pure and pellucid, in which Rutile in capillary and acicular crystals is thickly disseminated in every part. Cut, following the pyramid, in a conical shape. Exhibits internal fracture. Polished all over. 3 by 3 ins.

San Juan del Rey, Minas Geraes, Brazil.

No. 82. QUARTZ, pure pellucid, exhibiting interesting internal fracture and iridescence, with acicular crystals of Rutile. 3½ by 1¾ ins.

San Juan del Rey, Minas Geraes, Brazil.

No. 83. QUARTZ, very limpid and white, containing needle-like crystals of Rutile in parallel bunches. "Love's darts." Exhibiting fracture and iridescence. 3½ by 2 ins.

San Juan del Rey, Minas Geraes, Brazil.

No. 84. QUARTZ, interspersed with extremely fine reticulated capillary crystals of Rutile, the Oxide of Titanium. Polished all over. 2½ by 1½ ins.

Minas Geraes, Brazil.

No. 85. Pure pellucid QUARTZ, containing an air bubble, also parallel acicular crystals of Rutile, with a white exterior, probably caused by air. 2 by 2 ins.

Minas Geraes, Brazil.

No. 86. SMOKY QUARTZ or "CAIRNGORM." The "*Mormorion*" or "*Morion*" of Pliny, containing parallel acicular crystals of Rutile. This is the true "Veneris crines"* of Pliny, who evidently met the Rutile in the smoky variety of Quartz. The term, "*Flèches d'amour*" subsequently has been applied to all the varieties containing the needle-like crystals. 2½ by 1½ ins.

Minas Geraes, Brazil.

---

* "*Veneris crinis nigerrimi nitoris continet speciem rufi crinis,*" *which corresponds precisely with this smoky variety containing the red hair-like crystals. Lib.* xxxvii., 69.

No. 87. Section of QUARTZ containing very fine acicular and capillary crystals of Rutile, thick at the base. Exhibits internal fracture. Polished both sides. 3¾ by 2½ ins.

Minas Geraes, Brazil.

No. 88. QUARTZ, colourless and pellucid, containing thin parallel crystals of Rutile. Beautifully iridescent. Cut and polished all over. 3¼ by 2¾ ins.

Minas Geraes, Brazil.

No. 89. QUARTZ, containing golden acicular crystals of Rutile (TiO_2), shooting in every direction. The *Chrysothrix* of the Orphic poem. White, feathery, polished all over. 3½ by 2 ins.

Minas Geraes, Brazil.

No. 90. QUARTZ SPECIMENS, two small. One oval, impregnated with short crystals of Göthite (after Göthe, the poet)—the hydrous Oxide of Iron. Cut and polished on both sides, with bevelled edge. Another, more pellucid, cut *en cabochon*, tallow-topped, also containing Göthite (hydrous Oxide of Iron). These specimens are from the "Beresford-Hope" Collection, formerly exhibited at South Kensington, but dispersed lately at "Christies'." About 1 in. long each.

Brazil, South America.

No. 91. QUARTZ, large fine slab, pure pellucid, enclosing long comparatively thick well-defined black crystals of Tourmaline (from a Cinghalese name, *Turamali*) running in various directions. Shows internal fracture with cloud-like feathers. Transparent, cut and polished. 9 by 5½ ins.

Madagascar.

This Quartz from Madagascar is used, like the Brazilian "Pebbles,' largely for optical purposes, but does not occur usually in "pebbles," but in large massive lumps. Tourmaline is a hydrous silicate of alumina, magnesia, and iron, with traces of lime, etc.

No. 92. QUARTZ, white, feathery, containing solid crystals of black Tourmaline, coated with ferruginous oxide, some being quite red. Transparent but feathery. 4¾ by 2¾ ins.

Matumbaugh Mts., Madagascar.

Coll. Arthur Wells.

No. 93. QUARTZ, pure and pellucid, enclosing short, thick, crystals of Tourmaline, some black, and others, the smaller, red. Polished on one face ; surface exhibits splintery fracture. 3½ by 3 ins.

Matumbaugh Mts., Madagascar.

No. 94. QUARTZ, pure, pellucid, containing thin acicular crystals of Tourmaline with a little green Chlorite. Iridescent, exhibits fracture on exterior, and impressions of other minerals. Transparent. 3½ by 3 ins.

Matumbaugh Mts., Madagascar.

No. 95. QUARTZ, transparent, containing short, thick crystal of olive green Tourmaline, partly embedded and partly on surface. 2¼ by 2 ins.

Minas Geraes, Brazil.

No. 96. QUARTZ, conical, colourless, enclosing numerous capillary crystals of black Schörl (Tourmaline); fractured internally. Polished all over. 2½ by 2 ins.

Minas Geraes, Brazil.

Schörl is an early name given to Tourmaline, first published as Schurl in Ercker's "*Aula Subterranea*" in 1595.

No. 97. GROUP OF QUARTZ CRYSTALS, scattered, containing capillary crystals of Göthite (hydrous Oxide of Iron). 3 by 2 ins.

Minas Geraes, Brazil.

No. 98. GROUP OF SMOKY QUARTZ CRYSTALS, containing micaceous specular iron on matrix of Hæmatite, from Aῖμα =blood, the iron ore of commerce. 3½ by 3 ins.

District of Cleator Moor, Cumberland.

No. 99. GROUP OF QUARTZ CRYSTALS, with micaceous specular iron on a base of Hæmatite. 3¼ by 2 ins.

Cleator Moor, Cumberland.

No. 100. QUARTZ, crystallized, impregnated with ferruginous oxide of iron, and quite red. 2 by 1½ ins.

Iron mines of Cleator Moor, Cumberland.

No. 101. QUARTZ, containing filaments of green Asbestos with Chlorite on the exterior. Well polished. 2¾ by 2 ins.

Barèges, Hautes Pyrenées, France.

No. 102. QUARTZ, containing vermicular crystals of Chlorite, polished on one side; the other side is crystalline, exhibiting parallel striæ, and made up of many crystals. 2¾ by 2¾ ins.

Canton Uri, Switzerland.

" Mawe " Collection.

No. 103. QUARTZ, containing Chlorite, which is disseminated on a plane in the interior. Pure transparent crystal. Polished all over. 2⅛ by 1¾ ins.

Minas Geraes, Brazil.

No. 104. GROUP OF ROCK CRYSTALS, transparent, enclosed and covered by green Chlorite. 3¼ by 2 ins.

Tavetschthal, Tyrol.

No. 105. SMOKY ROCK CRYSTAL, doubly terminated, transparent, enclosing a little Chlorite.   2⅛ by ¾ ins.

Maderanerthal, Uri, Switzerland.

No. 106. ROCK CRYSTAL, containing mossy-looking white dessicated Silica with Chlorite. A crystal of Quartz intrudes its pyramid at the base, which is coloured by iron. A very interesting and unusual crystal.   2½ by 2¼ ins.

Tavetschthal, Switzerland.

"Tennant" Collection.

No. 107. QUARTZ, small group of transparent crystals, thickly covered with Chlorite.   1¾ by 1⅜ ins.

Disentis, Switzerland.

No. 108. Specimen of pure pellucid QUARTZ enclosing Chlorite. Polished all over. Transparent.   5 by 3½ ins.

Minas Geraes, Brazil.

No. 109. ROCK CRYSTAL, enclosing prismatic crystals of Hornblende like acicular Actinolite. They are soft, and disintegrating into a variety of Asbestos. Calcite, in crystals, is on the outside, with some Rutile (Oxide of Titanium). 3½ by 1½ ins.

Madagascar.

No. 110. QUARTZ, perfectly pure and pellucid, enclosing long crystals of Epidote in transparent prismatic crystals of a pistachio-green, some terminated. Exhibits lovely iridescence through internal fractures. Polished all over.   6¾ by 3½ ins.

Minas Geraes, Brazil.

> Epidote is a hydrous silicate of alumina, lime, and iron, with sometimes traces of magnesia and manganese; and is often found impregnating Quartz, though not in such distinct and beautiful crystals as in the above specimen.

No. 111. Small doubly terminated QUARTZ crystals, white, distorted, containing Carbon. ¼ to about ½ an inch.

Lizzo, Bologna, Italy.

No. 112. QUARTZ CRYSTALS, minute, doubly terminated, containing carbonaceous matter—many distorted. About ⅛ in. long.

Lizzo, Bologna, Italy.

No. 113. QUARTZ CRYSTALS. Small in glass tube. A group, doubly terminated, containing carbonaceous matter.

Lizzo, Bologna, Italy.

No. 114. QUARTZ CRYSTAL, white, cloudy, with distorted apex, containing Amianthus (Amphibole) in thin crystals, associated with Mica. 2 by 1¼ ins.

St. Brida, Switzerland.

CROCIDOLITE.

THIS beautiful and extremely interesting mineral has long been known to mineralogists, and was described by Klaproth as early as 1811. Its name is derived from the obsolete word Κρόξ=wool, in allusion to its fibrous structure or woolly appearance. Hausmann gave this name to a fibrous mineral from South Africa, which was shown by Stromeyer to be a hydrated Silicate of Iron, Sodium, and Magnesium.* A similar substance, brought by a traveller named Lichenstein, had been examined by Klaproth and described under the name of "Blaueisenstein." The variety which is so well represented in the Derby Collection is called Faserquarz, or Faserkiesel, and is not the true Crocidolite, the latest work by MM. Renard and Klement describing it as a "mineral deposited between the fibres of the original Crocidolite from a Siliceous solution," and not, as supposed, that the Quartz was a pseudomorph, after fibrous Crocidolite. It is one of the minerals to which the term "Cat's-eye" is given, and for which gem it was sold upon its arrival at a very high price. In consequence of a beautiful fibrous reflection it possesses, caused by the seams or masses making an angle with the opposite surfaces of 106°, with distinct colours of gold and brown, it looks when cut " en cabochon," with a high ridge, and the longer diameter of the oval at right angles to the fibres or filaments, like a Cat's-eye, with a beautiful chatoyant line of light, presenting a most brilliant appearance.

It has been largely used for ornamental purposes, having been made into tables, vases, tazze, and innumerable other ornaments, as well as for bijouterie.

* " *Gotting, Gelehr.,*" *Aug.* 1831, 1585.

Cat's-eye.

No. 8. Cat's-eye.

No. 115. CAT'S-EYE. Katzenauge Germ.; Œuil de chat
Fr., made from Crocidolite. A fine specimen, with a good
golden chatoyant, luminous, mobile yellow ray, with dark
sides—produced by being cut en cabochon transversely across
the fibres. 3¼ by 2¼ ins.

Asbestos Mountains, Orange River.

No. 116. CROCIDOLITE, South African Cat's-eye, Tiger's-
eye, Faserquarz. A large thick fine block, wedge-shaped, of a
beautiful rich chatoyant golden-brown colour. Unusually
wide. Well polished on both sides. 8¼ by 5 ins.

Composition of this brown variety, (Tiger's-eye) by Klaproth
gave :

Silicon Dioxide	98·5
Ferric Oxide ...	1·5
	100·0

which is confirmed by Messrs. A. Renard and C. Klement,*
who made a very careful study of this mineral. Their
analysis yielded :

Silicon Dioxide	93·05
Ferric Oxide	4·95
	98·00

with simply traces of alumina, lime, magnesia, and a little
water. It differs but little from the blue Hawk's-eye variety,
the analysis of which is given in its proper place, No. 123.
Associated with Jasper and Magnetite in thin seams—more or
less contorted. This is the Faserquarz sometimes confounded
with the true Crocidolite. The examination, both microscopi-
cally and analytically, leads Messrs. Renard and Klement to
the conclusion, as Professor Rudler says, "that the Quartz is
not a pseudomorph after fibrous Crocidolite, but was deposited
between the fibres of the original Crocidolite from a siliceous
solution which permeated the mineral." The rocks in which it
has been discovered are undoubtedly altered.

Hardness 7·0. Specific gravity of this variety gave 3·05.
8¼ by 5¾ ins.

Occurs in the Quartzose Schists called the Asbestos
Mountains, 23° and 24° long. east of Greenwich.

Griqualand West, north of the Orange River, South Africa.

* "Bull. Ac. Roy. Bel.," 3, 8.

No. 117. CROCIDOLITE, Faserquarz, Tiger's-eye. Chatoyant golden-brown colour, exhibiting fibrous formation. 6¼ by 2⅞ ins.

Quartzose schists of the Asbestos Mountains, Orange River, South Africa.

No. 118. CROCIDOLITE, South African Cat's-eye. Chatoyant golden-brown colour. A solid block. 7 by 3 ins.

Asbestos Mountains, Orange River, South Africa.

No. 119. CROCIDOLITE, Faserquarz, South African Cat's-eye. Of a chatoyant golden-brown, associated with blue Crocidolite—*Falkenauge*, Hawk's-eye. 5 by 3⅝ ins.

Asbestos Mountains, Orange River, South Africa.

No. 120. CROCIDOLITE, Tiger's-eye. Of a chatoyant golden-brown colour, with cross bands of blue *Falkenauge*. 5½ by 2½ ins.

Asbestos Mountains, Orange River, South Africa.

No. 121. CROCIDOLITE, Faserquarz. Chatoyant golden-brown. Oval. Cut as a paper-weight. 3 by 2¼ ins.

Asbestos Mountains, Orange River, South Africa.

No. 122. CROCIDOLITE. Chatoyant golden-brown. Oval. Cut as a paper-weight. 2⅞ by 2 ins.

Asbestos Mountains, Orange River, South Africa.

No. 123. CROCIDOLITE. Blue variety known as *Falkenauge* (Hawk's-eye). Exhibits fibrous structure and great mobility. The composition of this variety according to MM. Renard and Klement is :

Silicon Dioxide	93·45
Ferric Oxide	2·41
Ferrous Oxide	1·43
	97·29

with traces of alumina, lime, magnesia, and a little water.

Hardness about 7·0, same as the yellow. Specific gravity, 2·69 in comparison to 3·05 of the yellow. 5 by 3 ins.

Asbestos Mountains, Orange River, South Africa.

No. 124. CROCIDOLITE, Faserquaz, Tiger's-eye. Golden-yellow, chatoyant, exhibiting fibrous formation. A large thin specimen. Polished on both sides. 8 by 5 ins.

Asbestos Mountains, North of the Orange River, South Africa.

Unusually wide specimen. The veins of Crocidolite do not run as a rule more than two to three inches wide, so that the above specimen and No. 116 are quite unusual. When of such width and large enough, vases are cut from the solid in one piece.

No. 125. Ball of bluish CROCIDOLITE. Chatoyant, exhibiting fibrous structure. Diameter 1½ in. Stained, cut, and polished at Oberstein.

Orange River, South Africa.

No. 126. Ball of CROCIDOLITE, stained of a reddish colour, chatoyant, exhibiting fibrous structure. Diameter, 1½ in. Cut and polished at Oberstein.

Orange River, South Africa.

No. 127. Ball of CROCIDOLITE, stained of a reddish colour. Diameter, 1½ in. bare. Cut and polished at Oberstein.

Orange River, South Africa.

No. 128. Ball of CROCIDOLITE, of a whitish blue, partly stained and then subjected to great heat. Diameter, 1½ in. full. Cut and polished at Oberstein.

Orange River, South Africa.

No. 129. Ball of CROCIDOLITE, of a whitish blue, faceted in equilateral triangles, stained and subjected to great heat. Diameter, 1½ in.

Orange River, South Africa.

No. 130. Large ball of golden-brown chatoyant CROCIDOLITE, exhibiting its asbestiform structure. Diameter, 2 ins. Cut and polished at Oberstein.

Asbestos Mountains, Orange River, South Africa.

E

AVENTURINE QUARTZ.

THERE are two varieties of *Aventurine Quartz*, one the common—occurring at Aventura, in Spain, from which the name is probably derived—and the other a rarer green variety from India, both of which are represented in this collection.

The common is a brownish-red Quartz, massive or translucent, which contains gold or brass-yellow glistening leaves of Mica, Hæmatite, or other mineral, which gives it its interesting glistening appearance. It is most probably the stone described by Pliny under the name of Sandaresos, or Sandastros, who, speaking of it, says: "*Commendatio summa, quod velut in translucido ignis obtentus, celantesque se transfulgent aureæ guttæ, semper in corpore, nunquam in cute,*" which would be a good description, as the Mica is *in* the stone and not only on the surface. It was greatly used for jewellery purposes in former years, but the manufacture of imitation. Aventurine in large quantities, to which the name "Goldstone" was given, and which is really more beautiful than the genuine (although being only a paste and easily broken), has caused its decline. The imitation is made largely in Bohemia and France, as well as at Venice, and is manufactured by throwing brass filings or powdered brass into colourless strass, or sometimes into a composition, which is found to be more easily worked, and consists of

105 parts of Quartz,
85 purified Potash,
230 Tin and Lead Alloy
50 Brass Powder.

The Italians make an artificial Aventurine out of Siliceous Oxide of Copper.

A magnificent vase of Siberian Aventurine was bequeathed by the late Sir Roderick Impey Murchison to the Museum of

Practical Geology, who had received it as a present from the Emperor Nicholas for his geological and mineralogical discoveries in Siberia.

The rare green Indian Aventurine, which has not been imitated, is coloured, it is generally stated, by being impregnated with Chlorite ; but the author suggests that the colouring property is more probably Fuchsite, a Chromium Mica. The first specimens were brought to this country by Col. C. S. Guthrie, magnificent examples of which have been bequeathed by him to the Indian Museum, as well as specimens to the British Museum. It is now cut into vases and ornaments at Simla, in India, many of which, being inlaid with precious stones and pure Indian gold, are greatly appreciated in the European markets.

No. 9. Aventurine.

No. 131. AVENTURINE QUARTZ. Sandastros, *Pliny.* Quartz spangled with Mica (reddish tinge), rectangular-cut specimen. $2\frac{3}{4}$ by $1\frac{1}{2}$ ins.

Aventura, Spain.

No. 132. AVENTURINE, green base, with beautiful spangles of mica, coloured probably by Fuchsite, a Chromium Mica. Polished on one side. Very rare. $4\frac{3}{4}$ by $4\frac{1}{2}$ ins.

India.

No. 10. Impure through distinct minerals impregnating the mass.

No. 133. EISENKIESEL (IRON FLINT) or QUARTZ CRYSTALS in aggregation, impregnated with hydrated Sesquioxide of Iron (Ferric Oxide).

An analysis of this compact red Quartz made by Schnabel from the same locality (Iserlohn) yielded—

Silica	...	...	...	94·93
Alumina	...	...	...	0·42
Ferric Oxide	...	...	3·93
Magnesia	...	...	...	0·73
			100·01

3 by 3 ins.

Iserlohn, Westphalia.

E 2

No. 134. EISENKIESEL (IRON FLINT) or QUARTZ CRYSTALS in aggregation. Red, impregnated with hydrated Ferric Oxide. 2½ by 2 ins.

Iserlohn, Westphalia.

No. 135. CRYSTALS of QUARTZ (seven), ferruginous, opaque, doubly terminated. *Hyacinth of Compostella.* Small, in glass-capped box.

San Jago di Compostella, Galicia, Spain.

No. 136. QUARTZ CRYSTALS (four), black, doubly terminated ; small, some distorted. ½ to ¾ ins.

Bologna, Italy.

No. 137. QUARTZ CRYSTALS (two), black, doubly terminated in dodecahedrons. About ½ in. long.

Porretta, Italy.

No. 11. *Containing liquids in cavities.*

No. 138. QUARTZ crystals, a small white group, containing a drop of water with an air bubble in pyramid of the larger crystal, exhibiting impressions of other crystals on the prism. 1¾ by ¼ in.

Porretta, Italy.

No. 139. QUARTZ, containing a drop of water with an air bubble. Distorted. Pure white. Exhibits horizontal striations on the prism. 1 by ⅝ in.

Porretta, Italy.

No. 140. QUARTZ, white, containing a drop of water with an air bubble. Doubly terminated. Feathery, cavernous and striated across the prism. 1 by 1 in.

Porretta, Italy.

No. 141. QUARTZ CRYSTAL, white, containing a drop of water with an air bubble. Transparent. Hacked in appearance. ⅞ by ½ in.

Porretta, Italy.

B. CRYPTO-CRYSTALLINE VARIETIES.

CHALCEDONY.

CHALCEDONY, which occurs white, yellow, brown and blue, may be termed a translucent or cloudy variety of amorphous and crystalline Quartz. It is the constituent of nearly all the stones used by the cameo and intaglio engravers, under the names of Sard or Carnelian, Onyx, etc. It was procured by the ancients from Egypt, who held it in the highest esteem. It is found generally massive, mammillary, botryoidal, globular, reniform, and stalactitic. The name is from Chalcedon, its ancient locality in Asia Minor. Pliny knew the stone, but does not describe it exactly, speaking of it partly as " Murrhina," for which see page 168.

It is found in cavities in many rocks, also in gangues, pebbles and boulders. Very fine specimens come from Iceland (see No. 142) and the Faroe Islands, as well as from Galgen-Berg; Oberstein on the Nahe, Cornwall, Scotland, Greenland, Hungary, Saxony, Ceylon and many parts of India and America. In Iceland it is very abundant, and is highly esteemed by the inhabitants, who attribute to it medicinal properties.

No. 1. *Chalcedony.*

CHALCEDONY. Chalcedon in Asia Minor, from whence it was first obtained. Murrhina,* Chalcedoine *Fr.;* Chalcedon *Germ.* " *Achates vix pellucida, nebulosa colore griseo mixta.*"—

* *Lib., Pliny*, xxxvii., 7.

Wall 83, 1747. An amorphous variety of Silica. Grey, white, pale and dark brown, black, and often of a delicate blue. Massive, mammillary, botryoidal, stalactitic, and often filling cavities in rocks.

Hardness 7·0. Specific gravity 2·6 to 2·64.

No. 142. MAMMILLATED CHALCEDONY, of a pale bluish colour with a waxy lustre. It has evidently formed the lining of a cavity in a rock. A very characteristic specimen, the whole mass being pure Chalcedony. 4½ by 3 ins.

Reykjavik, Iceland.

No. 143. Slab of CHALCEDONY, in onyx-like circles. White. and greyish blue. Polished on both sides, bevelled. 5½ by 3 ins.

Cambay, India.

Coll. Arthur Wells.

No. 144. CHALCEDONY, oval slab, bluish grey to white. Polished all over, including sides, at Oberstein, Germany. 4¾ by 3¾ ins.

Uruguay, South America.

No. 145. CHALCEDONY, oval slab with a pale amethystine tint in cloud-like streaks or strata. Bevelled edge. 4½ by 3 ins.

Reykjavik, Iceland.

Coll. Dr. Tayler.

No. 146. Thin section of CHALCEDONY, of a bluish-white. Translucent. Polished on one side. 4¼ by 2¾ ins.

Reykjavik, Iceland.

No. 147. CHALCEDONY, pure, nearly square. By transmitted light shows mammillated structure. Polished both sides. 3½ by 3 ins.

Kangertluar Fiord, West Greenland.

From Dr. Tayler's Greenland Expedition.

No. 148. CHALCEDONY, brownish-white, slightly tinged with iron. Thick specimen. 3¼ by 1¼ by 1 in. thick. Polished all over.

Kangertluar Fiord, West Greenland.

Coll. Dr. Tayler (Greenland Expedition).

No. 149. FORTIFICATION CHALCEDONY, consisting of minute concentric strata. Very fine, white in the centre. Translucent. Polished both sides. Bevelled edges. 2⅞ by 1⅝ in.

Meerut, India.

Coll. Arthur Wells.

No. 150. CHALCEDONY, bluish-grey with white streak. Translucent. Octagonal. 2¼ by 2½ ins.

Kangertluar Fiord, West Greenland.

Coll. Dr. Tayler (Greenland Expedition).

No. 151. CHALCEDONY, reddish-blue banded with amorphous Quartz. Translucent. 2¼ by 1¾ ins.

Minas Geraes, Brazil.

No. 152. Oval specimen of CHALCECONY, bluish-white, exhibiting mammillated structure by transmitted light. Cut and polished, with bevelled edges. 2 by 1½ ins.

Cambay District, India.

No. 153. Translucent wax CHALCEDONY, *Cerachates* of Pliny. Fawn-grey, with portion of an Agate in the centre. A thick translucent specimen. 4¼ by 3 ins.

Near Reykjavik, Iceland.

No. 154. CHALCEDONY, oval, with brown Jasper in centre. Translucent. Bevelled edges. Polished all over. 3 by 1⅝ ins.

Cambay District, India.

No. 155. CHALCEDONY, reddish tinge, oval specimen. Exhibiting mammillated structure by transmitted light. Cut and polished all over. 4 by 3¼ ins.

Cut at Oberstein on the Nahe, but from Brazil.

No. 156. CHALCEDONY, prettily marked with brown. Translucent. Irregular oval, bevelled edge. Cut and polished all over. 3½ by 2¾ ins.

Meerut district, India.

"Mawe" Collection.

No. 157. Ball of pure CHALCEDONY of a bluish and pinkish-white. Well polished. Diameter 1½ ins.

Cut and drilled at Oberstein, but from Brazil.

No. 158. Ball of CHALCEDONY, bluish-grey with artificially coloured fortification of red Sardonyx. Diameter 1½ ins. bare. Cut and drilled at Oberstein.

District of Minas Geraes, Brazil.

No. 159. Ball of pure CHALCEDONY, with concentric layers or strata of faint-coloured Sardonyx at top. Diameter 1¾ ins. Cut and drilled at Oberstein.

Brazil.

No. 160. Ball of pure CHALCEDONY of a buff and bluish-white colour. Diameter 1½ ins. Cut, drilled, and polished at Oberstein.

Uruguay, South America.

No. 161. Ball of CHALCEDONY stained a pinkish hue. Diameter 1 in. full.

Cut and drilled at Oberstein, but from South America.

No. 162. Ball of brown CHALCEDONY with alternate layers of brown, red, and bluish tints. Diameter 1½ in.

Cut and drilled at Oberstein, but from South America.

SARD OR CARNELIAN.

SARD, or Carnelian, is a translucent horn-like variety of Quartz, occurring yellow, brown, and red. It was known to the ancients under the name of Sarda, from the Arabic meaning yellow. The two varieties are particularly prized by jewellers and lapidaries. Those having a pale colour or yellow tinge are called Sard, and those being of a dark red colour, Carnelian. The latter is called by the French, "*Cornalines de la vielle roche.*" It is found at all the usual localities of Agate, and is employed largely by jewellers, as well as seal engravers.

The Romans carved greatly in Sard or Carnelian. The seal of Michael Angelo, valued at 50,000 francs and said to have been engraved by Maria de Descias, after the original of Praxiteles, the bust of Ulysses, Hercules killing Diomede, Jupiter, Mars and Mercury, all in the Imperial Library at Paris, are cut out of Carnelian.

————

Sard is the Greek Σάρδιον = the Sard of *Theophrastus*, and the *Sarda* of Pliny, who mentions it several times. Carnelian, from the Italian *Carniola* (Latin, *Carocarnis*), signifying flesh, is the same mineral. *Carneol*, Agricola, "*Agates fere pellucida, colore rubescente,*" *Wallerius*, 82, 1747 : Cornaline or Sardoine, *French*. A clear brown and red Chalcedony of a fine rich colour by transmitted light. The best coloured Carnelians, according to M. Gauthier de Claubry, contain organic matter which gives off carbonic acid when heated with oxide of copper. Specific gravity 2·59 to 2·63. An analysis made by Bindheim of a specimen of Carnelian yielded—

Silica	..	... ...	94·00
Alumina ...		... ...	3·50
Oxide of Iron		... ...	0·75
			98·25

No. 2. Sard or Carnelian.

No. 163. SARD or CARNELIAN (L., *caro-carnis* - flesh ; and σάρξ=flesh), a clear red Chalcedony with the peculiar waxy nature of this variety of Silica. Reddish-brown by transmitted light. Polished on one side. 4 by 3 ins.

Surat, India.

No. 164. SARD or CARNELIAN, an interesting specimen, reddish-brown. Exterior of pebble exhibited. 4 by 3¾ ins.

Meerut District, India.

No. 165. Oval disc of CARNELIAN. The colour is artificial. Polished on both sides and on the edge. 4¾ by 4 ins.

Cut at Oberstein on the Nahe, but from Brazil.

A good example of artificial colouring.

No. 166. Rectangular specimen of CARNELIAN, of a beautiful clear red colour with faint onyx-like markings. Colour artificially deepened. Cut and polished. 4 by 2½ ins.

Meerut, India.

No. 167. CARNELIAN PEBBLE, portion of. In concentric layers with band of white Chalcedony. Exhibits exterior. One side polished. 2½ by 2¼ ins.

Meerut, India.

No. 168. CARNELIAN, slab of. Showing mammillated form by transmitted light. Polished all over.   2¾ by 2½ ins.

Minas Geraes, Brazil.

No. 169. CARNELIAN, flesh-coloured.   Square, with Chalcedony.   Translucent   Polished both sides.   2½ by 2¾ ins.

District of Mocha, Arabia.

No. 170. SARD or CARNELIAN, cut from portion of a pebble exhibiting exterior, with a centre of crypto-crystalline Silica (irregular with an amethystine tinge.   2 by 2 ins.

District of Mocha, Arabia.

No. 171. Fine specimen of SARD or CARNELIAN, in mammillations.   Reddish-brown by transmitted light.   Centre of silicified wood.   Translucent.   Polished both sides.   5 by 3 ins.

District of Mocha, Arabia.

No. 172. CARNELIAN, red in onyx-like formation, irregular. Translucent.   Polished on one side.   3¼ by 2¾ ins.

District of Mocha, Arabia.

No. 173. CARNELIAN, bright vivid red surrounding transparent Chalcedony of a whitish-violet tinge.   Small but very striking.   2½ by 2 ins.

Cambay, India.

No. 174. CARNELIAN, vivid red surrounding Chalcedony (transparent) of a violet tinge.   Small but striking.   2½ by 2 ins.

Surat, India.

CHRYSOPRASE.

THIS beautiful variety of Silica, occurring always massive, is of an apple-green, grass-green, and whitish-green colour, and derives its name from χρυσός = golden and πράσον a leek. A stone was known to the ancients under the name Chrysoprasus, but it is not the Chrysoprase of to-day. It was discovered in 1740, by a Prussian officer, in Silesia, where it is looked upon with great superstition, the common Silesians wearing it round the neck as a charm against illness and pains. It is used for ornamental purposes and mosaic work, its colour blending well with that of other stones ; also for jewellery purposes—latterly having become quite fashionable, although nearly all the specimens now being sold for Chrysoprase are simply white Chalcedony artificially coloured. It is a Silica with a little carbonate of lime, alumina, and oxide of iron and nickel, to the latter of which it probably owes its colour. It is a difficult mineral to cut well, as it easily splinters, the greatest care being necessary, particularly in faceting it. It has to be polished on a tin plate with rotten stone, care being taken to prevent it getting too hot, as it would become opaque and turn grey, losing its beautiful colour. Imitations are made of it by mixing 1,000 parts of strass with 5 parts of oxide of iron and 8 parts of oxide of nickel.

No. 3. Chrysoprase.

CHRYSOPRASE, an apple-green Chalcedony, the colour being due to a small percentage of oxide of nickel. Klaproth found 1·0 per cent. NiO., and Rammelsberg 0·41 per cent. NiO. Splintery.

An analysis made by Klaproth yielded :

Silica...	...	...	...	...	96·16
Oxide of Nickel	...	...	...		1·00
Lime, Magnesia, and Ferrous Oxide.					Traces

97·16

175. CHRYSOPRASE, χρυσός golden, and πράσον - a leek. Apple-green colour in two tints. The beautiful green is caused by the presence of 0·40 to one per cent. of the oxide of nickel. Cavities at the back showing coralline structure. Polished on one side. 6 by 3½ ins.

Kosemütz, Silesia.

No. 176. CHRYSOPRASE, a beautiful apple-green colour of two tints, caused by oxide of nickel. Cavities at the back showing coralline structure. 5¾ by 3¼ ins.

Kosemütz, Silesia.

No. 177. CHRYSOPRASE, with a cavity in centre, probably formed by a coral. Good apple-green colour. Polished on one side. 3½ by 2¼ ins.

Kosemütz, Silesia.

No. 178. CHRYSOPRASE, pale apple-green colour, but deeper at one side than the other. Good colour at side of specimen. Convex at top. Well polished. 3½ by 2½ ins.

Baumgarten, Frankenstein, Silesia.

Mawe Coll.

PRASE.

PRASE is a translucent spotted Quartz of a muddy olive-green-leek colour, deriving its name from πράσον = a leek, and is not a stone of much value, although of great interest. Pliny mentions it several times, but it is doubtful whether he alluded to the Prase of the present day; more probably his remarks were alluding to the massive Emerald. " *Vilioris est turbæ Prasius,*" Pliny. It was also a name applied to a crystallized Quartz. An analysis made by Beudant of a specimen of Prase yielded :

Silica	...	... 95˙25
Alumina	...	... 0˙41
Ferrous Oxide		... 2˙66
Lime	...	... 1˙00
Magnesia	...	... 0˙67
		99˙99

Conchoidal fracture. Vitreous resinous lustre. Specific gravity 2˙66 to 2˙68. It contains besides Silicon a little alumina, oxide of iron with lime and magnesia. It is found in India and many other parts of Europe.

No. 4. Prase.

No. 179. PRASE, a dull green opaque variety of Quartz in parallel irregular lines interspersed with white. Wedge shaped. Polished on one side. 2½ by 2⅛ ins.

Breintenbrünn, Zwikau, Saxony.

PLASMA.

PLASMA, from πλάσμα = image, is of a bright leek, and occasionally emerald-green, variety of Chalcedony. Pliny mentions it under the Jaspers (Iaspis xxxvii., 37). It is flinty, translucent, and seldom free from spots or black dots consisting of pyrites, with sometimes pale patches produced by its porosity. It was used by the ancients, many specimens being found in the ruins of Rome, and it is found in many parts of India, and at Baumgarten in Silesia.

No. 5. *Plasma.*

No. 180. PLASMA, leek-green with a little white. Translucent. Polished on one side. 2¼ by 2¼ ins.

Banda, India.

No. 181. PLASMA, pretty bright green with small white spots surrounding flammeate red Jasper (blood colour). 2½ by 1¼ ins.

Banda, India.

No. 182. PLASMA, bright green with white spots surrounded by flammeate red Jasper. The external very greasy looking. 2¼ by 1¼ ins.

Banda, India.

No. 183. PLASMA, leek-green, spotted with a greasy-white Silica. Rough exterior. Polished on one side. 3 by 3 ins.

Banda, India.

HELIOTROPE OR BLOODSTONE.

HELIOTROPE—derived from ἥλιος = the sun and τροπή = a turning, from the supposition that if the stone be immersed in water it changed the image of the sun into a blood-red—is a mixture of Chalcedony with earthy chlorite, and is well known on account of its small spots, often pisolitic in character, looking like blood. It was the *Heliotropium* of Pliny, who describes it " *Heliotropium nascitur in Æthiopia, Africa, Cypro, porracei coloris, sanguineis venis distincta.*" * It was highly esteemed in the Middle Ages by the superstitious, who regarded the red spots as the blood of Christ issuing from the stone. Many figures of Christ attached to the cross in Heliotrope were made, the cutting of the figure being so arranged that the red Jasper spots occurred on the breasts and hands where blood would be flowing from the wounds. A specimen of this description is in the Louvre, and is very highly prized. Bloodstone is extensively used in jewellery, particularly for signet rings and seals, being easily engraved and taking a high polish. It is, however, of little intrinsic value, many boats from India a few years ago bringing it to this country as ballast. The ordinary variety is opaque, but a rarer, associated with Jasper, and under that name, is quite translucent, such as the remarkable specimens Nos. 494 and 495 in the collection, which are exceptional. It is really the same stone as Plasma.

The word Bloodstone must not be confounded with the German " Blutstein," sometimes called in commerce Bloodstone. Blutstein is Hæmatite, the principal iron ore of commerce.

* *Lib.* xxxvii., 60.

HELIOTROPE or BLOODSTONE, a Chalcedony with earthy chlorite, occurs massive and in obtuse-angular lumps. Conchoidal fracture. Hardness 7·0. Specific gravity 2·61 to 2·63.

Variety of Plasma.

No. 184. HELIOTROPE or BLOODSTONE, fine green characteristic specimen with spots of Jasper coloured by red oxide of iron, looking like spots of blood. One side very red. Rectangular. Mounted in silver. Well polished. 5¾ by 3⅜ ins.

Banda, India.

No. 185. HELIOTROPE or BLOODSTONE, slab of. Green, sparely disseminated with red spots of Jasper and white Silica. Bevelled. Polished all over. 4¼ by 2⅜ ins.

Banda, India.

No. 186. HELIOTROPE or BLOODSTONE, green, of a so-called pisolitic texture (Professor Ruskin) with flammeate red. Square, with bevelled corners and edges. 3⅝ by 3⅜ ins.

Banda, India.

No. 187. HELIOTROPE or BLOODSTONE, green, of a so-called pisolitic texture (Professor Ruskin) with flammeate red. Square, with bevelled corners and edges. 3¾ by 3⅜ ins.

Banda, India.

No. 188. HELIOTROPE. Oval specimen, light and dark green tints, with patches of pisolitic red spots of Jasper, looking like blood. 4¼ by ¾ ins.

Cut and polished all over at Oberstein, but from India.

No. 189. HELIOTROPE or BLOODSTONE, half red and half green; light green at one corner. Granular or pisolitic. Bevelled edge. Polished all over. 3¼ by 2⅜ ins.

Banda, India.

Coll. John Luff.

F

No. 190. HELIOTROPE or BLOODSTONE dark green with a few red spots of Jasper. A solid block polished on one side. 4 by 2 ins.

Banda, India.

No. 191. HELIOTROPE or BLOODSTONE, dark green. with red spots. Opaque and cloudy. Bevelled edge. Polished. all over. 2¼ by 1⅝ ins.

Banda, India.

AGATE.

AGATE is a variety of quartz assuming varied and beautiful forms by the aggregation of amorphous, crystalline, and crypto-crystalline silica, in alternate zones or layers, the result of the deposition of silica either in cavities of a vesicular rock like that of the Galgen-Berg at Oberstein on the Nahe, or by forming veins, or accumulating in fissures or crevices, when it is of a brecciated description. How these agates, which occur in the basic igneous rocks, essentially plagioclase and augite rocks, are formed in the amygdaloidal cavities at Oberstein, has been a great question in the mineralogical world for years. Many thousands of specimens have passed through the hands of the author, exhibiting the varied points in the discussion of the theories and hypotheses advanced at different periods as to their formation. The rock in which the agates are found at Oberstein is a Melaphyre distinctly vesicular in character, the agates deposited in it assuming the shape and original form of the cavities, some of which are globular or round, while others are elongated and flattened, termed amygdaloidal or almond-shaped. The introduction of silica into these cavities (said to take place by tubular orifices which are often observed), forming agates, can easily be accounted for by the supposition that, at a certain period, these rocks were saturated by water or a solution highly charged with silica, which, whilst in a boiling state, percolating through and decomposing them, would fill the cavities, somewhat similar to the silicated solutions which have completely enveloped and permeated the forests of Arizona and Antigua, where complete trees are found agatized.

F 2

The solution might fill the cavities from all sides, or filter through by one inlet, which would account for the "inlets of infiltration" so often met with in agates, whilst the cooling of the solution—it being presumably in a heated state—commencing at the exterior or outer walls of the cavities, would account for the regular deposition of the silica in bands or concentric zones, which would continue until the whole of the silica would be exhausted. At times, perhaps, the gas which would be naturally generated in the centre of the cavity would become too strong for the already semi-solidified zones, and would force its way through from the centre to the outside rock, accounting for the "channels of egress" which are often met with, they being of an eruptive instead of irruptive character. The solution of silica might not be in a pure condition, if even it, as supposed, was in a fluid state, spreading over the portion of a country, but may have been a solution holding carbonic acid gas, which has, by the decomposition of the rocks through which it has passed, finally deposited free silica. The greater the alteration of the rocks are in appearance by decomposition, so it has been noticed the finer are the agates. One of the constituents of the rock which would be altered without difficulty would be the feldspar Labradorite, one of the ingredients of the Melaphyre of Oberstein,[*] which, acted upon by carbonated waters, the silicate of calcium would decompose and form a carbonate, setting free the silica. If the amount of silica entering a cavity were not sufficient to fill it, less ensuing by evaporation, the centre of the agate would be a cavity, so often observed and nearly always in a drusy, semi-crystalline or perfect crystalline state not infrequently of the variety Amethyst, with its violet-blue colour.

These crystals, it will be noticed, have always the apices of the pyramids directed towards the centre cavity from without —inwards, being in structure what is termed *endogenous.* Re-

* *Delesse — Sur le Porphyre Amygdaloïde d'Oberstein,* "*Ann. des Mines*" (4) xvi., *p.* 511.

markable examples of this will be found in the large sections, Nos. 216 to 218, which distinctly show by transmitted light, although the amethyst crystals are not in the very interior of the stone, that all the apices point to the interior.

The cavities of nearly all the rocks which, it is concluded, originally contained gas or steam-bubbles, extended to an amygdaloidal or other shape by a flow of lava when in a viscous or semi-solidified state, contain silica in one form or another. Even if the solution saturating the rocks were simply water, it has been well established by that eminent authority, Professor Daubrée, Director of the School of Mines in Paris, that at a high temperature, and under pressure, certain silicates become decomposed, depositing separate silica.

Some authorities believe that even cold water ceaselessly percolating through the rocks, decomposing them, would be sufficient for the formation of agates, but the author cannot agree with this for the same reason given by Bischof, who reckoned that to complete even the deposition of a single layer would require twenty-one years, whilst to form an amethyst of only one pound in weight it would be necessary for ten thousand pounds of water to enter the cavity and become evaporated, a process of such length of time that it is computed it would occupy 1,296,000 years.* And this for a stone only a pound in weight, whereas agates have been found of gigantic size, single specimens weighing at times one hundred-weight each. The hypothesis of Professor Haidinger submits that the formation or genesis of an agate may be better understood by supposing that the silica, instead of being *infiltrated* into the cavity, entered wholly through the walls of the cavity, so that both top and bottom, roof and base, would become simultaneously and uniformly coated with silica. But if this be correct, how could the solution, after the first and presumably outside layer were formed, gain access to the cavity? Again, if it were possible, would not the whole of the zones be of uniform colour instead of the varied tints they now assume?

If the deposition, on the other hand, takes place slowly by infiltration, how can we account for the beautiful fine zones being of quite the same thickness in many agates both at the top, bottom, and sides? We say many agates because other agates are met with which have flat horizontal strata, forming as it were a base, and pointing to the supposition that they are formed by a higher gravity of the silica, whilst the top and sides consist of circular concentric zones, perhaps of a lighter density. See Nos. 245 and 246 which exhibit a parallel floor, called by Professor Ruskin "lake-like," with concentric layers.

Bischof suggests that the horizontal bands were formed from a *rapid* introduction of silica, whilst the concentric layers infiltrated in *slowly*, spreading gradually over the walls without touching the floor.

"Agates, I think," says Professor Ruskin, "confess most of their past history," to which statement most mineralogists will agree, but it is doubtful if anybody yet has been able to correctly read that history. There are enigmas in connection with agates which their formation hardly elucidates. For instance, it is quite common on the one stone to have silica crystallized, crypto-crystalline, and amorphous in the shape of various zones or strata, which strata will vary in texture, some being more porous than others, in hardness, in transparency— a stone often being transparent, translucent, and opaque in a few inches of ground and in colour—bands changing from a milky-white to a bluish-grey, and from a deep red to a delicate rose-pink. But what has caused this difference? and why should a stone in one part be harder than another, its texture and its colour undergoing a change? Often also lines of agate are perfectly straight and parallel with each other on one part of the circle, but at another they dip and are tilted! General answers may be given to these questions, such as that the colour is changed by oxides of iron, manganese, nickel, etc., being met with, mixing and infiltrated, with the portion of silica it encounters and dyes, into the agate. But there is no absolute answer to these queries which time may perhaps unravel. It is deeply to be regretted that Professor John Ruskin, who has studied this great question and formed one of the finest collections known of agates, has not been able, through illness and advancing years, to attack the question thoroughly of the *formation* of agates. He has arranged, and, with all that mastery of language which he possesses, described a most instructive series of specimens of silica in the British Museum, replete with valuable suggestions, but not forming any definite hypothesis. The most remarkable study of the structure of the agate is the extraordinary and truly marvellous minuteness of the deposition of its zones or layers. These are deposited with wonderful regularity and delicacy, zone upon zone, often concentric, following with a minuteness which baffles description every irregularity of surface, each stratum, however minute, being distinctly discernible, if not to the eye, with the aid of a magnifying glass. So minute are many of these that Sir David Brewster, trying to measure them, met with the greatest difficulty, finally, however, giving as a result that some of the layers were $\frac{1}{17230}$ and $\frac{1}{33160}$ part of an inch in thickness.* See specimens Nos. 224, 226, and 228, which when carefully examined will be seen to consist of the most minute and regular zones.

The objection to the theory of Haidinger that the silica entered wholly through the walls of the cavity, has been pointed out, as well as that of slow infiltration. Perhaps the most reasonable elucidation of the question is that the deposition of silica took place from a *boiling* solution, which entered or infiltrated into the cavity neither too quickly nor at too slow a rate, by what is termed one or many "inlets of infiltration," that it commenced to cool at once, solidifying at the outer edge, the distinct zones being formed at each appre-

* "*Phil. Mag.*," *Vol.* xxii., 213.

ciable change of decreasing temperature until the gas in the interior had evaporated entirely, or become so compressed as to be strong enough to force its way through the semi-solidified mass and cause a "channel of egress." That both "inlets of infiltration" and "channels of egress" exist, though the latter may often have been in the first case an "inlet" which, offering the least resistance to escaping gas or solution, was turned into a "channel of egress," supposing the solutions, as suggested by Dr. Forster Heddle, were of different densities outside and within the cavity, pressure being exerted outwards against the semi-solidified or gelatinous silica. That occasionally—the solution being of a greater density at one time than another, silica ranging in specific gravity from 2·5 to even, in coloured varieties, 2·8—would run into a cavity with a flat bottom, and solidify quickly in strata generally as observed of greater thickness than any other portion of the agate, whilst the concentric zones, perhaps of a lighter and purer silica, would slowly arrange themselves above the flat layers, which would constitute their base. See specimens Nos. 245 and 246.

These remarks refer only to agates of a globular or amygdaloidal shape formed in cavities of vesicular rocks, and not to those of a brecciated description, such as are found at Schlottwitz in Saxony, whose deposition is effected in fissures forming complete siliceous veins; nor do they of course in any way apply to the siliceous pebbles forming the quartz conglomerate, like the specimens from Hertfordshire (see No. 522), or to those agates which occur in many of the cavities of stratified rocks.

On the exterior of many agates, and also forming in a film the first lining of the cavities is a ferruginous chlorite called *Delessite*, a hydrous silicate of aluminium and iron, with magnesium and lime, of an olive to blackish-green colour, probably formed through the decomposition of augite. It really constitutes a thin rind, and is probably the result of "general percolation, and not of local deposition." It sometimes intrudes itself into the agate, going completely through it to the centre disarranging and upheaving the strata in a remarkable manner. This is well illustrated by the interesting specimen No. 194 in the descriptive catalogue.

A somewhat similar mineral is *Chlorophæite*, a hydrous silicate of iron and magnesia, which coats the geodes, fissures or amygdaloidal cavities.

Agates, when removed by disintegration from their original rocks, are found as nodules in the beds of rivers, or scattered in the shape of pebble drifts. Scotland used to supply large quantities, principally from Montrose in Forfarshire, and from Perthshire at the Hill of Kinnoul, whilst the Cheviot agates come from the Northumbrian streams. India supplied great quantities of agates, formerly factories existing at Cambay, Broach, and Ratanpur, procuring their stones from the trap rocks of the Deccan and Rajmahal Hill, or from the agate-bearing gravels of Rajpipla, which are mined systematically.* Fine examples also come from Arabia. Agates are found, more or less, round the British coast, particularly at Scarborough, Brighton, Aberystwith, and the Isle of Wight; at Oberstein on the Nahe, Idar, Bohemia, Saxony, France, and many other parts of Europe; in many parts of South America, especially Uruguay, Paraguay, and Brazil; in Queensland (Agate Creek), New South Wales, and other Australasian localities.

The beautiful hard-looking *jaspers* in their many varieties of colour, red, brown, blue, and black, including the "riband," or striped jasper, which exhibits broad parallel bands, as well as the brown Egyptian conoidic forms, follow in order after the agates, the striped varieties being met with in Siberia, Saxony, and Devonshire; the large boulders of red come from Siberia and India, and small pebbles of nearly a blood-red colour are found strewn by thousands over the plains of Argos, whilst yellow pebbles come from Vourla, in the Bay of Smyrna, associated with other silicas. The pseudomorphous quartz, or

"*Man. Geol. India,*" *Pt.* iii. (*V. Ball*) 1881, *page* 503, *part* iv. *F. R. Mallett,* 147.

quartz under the form of other minerals which it has assumed through the replacement or alteration of those species, are of exceptional interest, one of the rarest of which, although a small specimen, is represented in the collection—HAYTORITE (see No. 525), a pseudomorph after Datholite, a borosilicate of lime, the series containing as well "Babel" quartz, so called because it has the appearance of a tower, having upon its under surface impressions of cubical fluor spar by its deposition over the crystals.

Silicified wood is Quartz, pseudomorphous after wood, retaining the original structure of the spores and bark by the deposition of silica from its own solution into the cells of the wood, finally enveloping and taking the place of the walls of the cells and bark, until only silica is left, the wood entirely disappearing. These, with the exception of a graphic granite, so named as it looks like Hebrew characters, and is sometimes called Hebrew stone, named Pegmatyte, quartz arranged in a parallel position in feldspar—conclude the crypto-crystalline or massive flint-like varieties.

It may be safely stated that all the true varieties of agate and jasper are represented in the "Derby" Collection in every form and size : some representing solid agates, with their interesting coatings of delessite, or ferric hydrate, in opaque, round nodules, or amygdaloidal semi-pebbles ; others, the thin crypto-crystalline sections, whose transparency exhibit marvellous depositions of silica in their banded or extremely minute concentric zones, or exhibiting the "inlets of infiltration" and "channels of egress," which assist to elucidate the history of the genesis of the agate ; whilst others display the beautiful varied colours and delicate natural tints for which they are so deservedly celebrated.

OBERSTEIN.

With an account of the manner of Cutting, Polishing, and Staining Agates.

OBERSTEIN is a small picturesque village on the Nahe, a tributary of the Rhine at Bingen, and has become the centre of the agate trade through the discovery, centuries ago, of agates in quantities, occurring in the amygdaloidal Melaphyre of the Galgen-Berg near Idar—a small village where the trap rock begins, and whose river, the Idar Bach, flows into the Nahe, furnishing Oberstein with immense water power, which is divided and subdivided into innumerable channels, turning hundreds of mills, at a nominal cost, which elsewhere have to be driven by steam power.

The principal rocks of Oberstein occur in three strata, the top bed being composed of a conglomerate of Quartz, Porphyritic and other stones worn into pebbles, and united by a ferruginous earthy material burnt by the second stratum, which consists of an amygdaloid trap, interspersed here and there with crystallized calcite. At the junction of these two strata in the mountain facing Oberstein is an enormous cavern, partly natural and partly excavated, which, by the aid of doors and windows, has been made into the village church.

The trap rock, which is of a vescicular character, is an olive-green colour running into brown, and extends for about two miles close by the riverside up to the small village of Idar, where another rock—a third variety—begins, forming the large hill Galgen-Berg, which produces the agates and chalcedonies which has rendered Oberstein and the district so celebrated. The specimens occur in nodules or masses in the rock, which is of darker or denser trap, of a basaltic nature. At many other places in the surrounding district agates, principally in nodules and geodes are found, although some are of a stalagmitic nature.

The Oberstein agates were for a long time justly prized, but there were distinct indications in 1827, to the great alarm of the inhabitants, of the supply beginning to fail, when happily the importation of specimens by thousands took place from Brazil, which supplanted those from the district—but principally through their cheapness and size, not quality—as the old specimens from Galgen-Berg, particularly the white, are better than either the Indian, generally termed Oriental, or Brazilian ; and here, the author particularly draws attention to the fact that he has purposely omitted the use of the word " Oriental " in the whole of this work ; for most of the specimens termed by the dealers and general public " Oriental " are really Brazilian or other South American or European stones dyed or deepened in colour. Certainly the Brazilian and Indian agates occur of a beautiful yellowish-brown, or natural orange, which Oberstein does not yield, and which the Germans have not yet succeeded in imitating ; but the word Oriental is nowadays used so recklessly—every dark specimen being termed " Oriental," irrespective of its locality —that the author has purposely avoided the use of the word.

The agate trade at Oberstein is divided into two branches, one devoted to cutting and polishing stones, the other to boring them. The cutting is performed by large mills, turned by water ; the grinding-stones, brought from Kaisers-lautern, near Mannheim, being from five to six feet in diameter, and made of a red sandstone grit, which are fastened upon a shaft which causes them to work vertically, rotating at about three revolutions per minute, and moistened continually by a stream of water ; the width of each stone is from a foot to fifteen inches, its surface possessing channels of various shapes corresponding to the forms of the objects desired. The workman has to lie horizontally upon a kind of stool, fashioned so as to fit the body, called a 'cuirass,' and then presses the agate against the mill, obtaining purchase by placing his feet against a block or pole attached to the ground. The mill-stone often becomes smooth by continual friction, but is made rough again by being struck with a sharp hammer, either (according to the variety of work required) fine or coarse. The polishing is effected by means of a fine clay, or rouge, and water, chiefly on wooden cylinders with tripoli, although soft metallic discs are sometimes used. It is, however, generally of a very inferior standard, the power of many scores of mills being so divided as to render a very high polish impossible.

The boring of agates is quite another branch, and is performed by means of the diamond, in the shape of *boart* and *carbonado*. Now, not only agates, but specimens of every description are cut at Oberstein and Idar, including every variety of silica, such as jaspers, siliceous corals, as well as crocidolite, malachite, lapis-lazuli, etc., and all the decorative minerals.

Another locality which is showing signs of great development is Waldkirch, in the Black Forest.

Periodically, sales by auction take place at Oberstein of Brazilian agates, which are imported from that country, when the great dealers have to show their judgment in purchasing lots that contain agates which are translucent and *solid to the core*, as a cavity, which more often than not occurs in the interior, is fatal to the manufacture of an important work of art out of the stone. The difference and beauty of the layers in an agate regulates its market value ; it will not be surprising, therefore, to learn that the artificial colouring of stones has been practised from the earliest times—even as far back as Pliny the process of darkening agates was known, and is imperfectly described by him in "Historiæ Naturalis" xxxvii., c. 75, as taking place in Arabia. Prof. Nöggerath, whose fine collection of agates, mostly from the melaphyre of Oberstein and the

Galgen-Berg, are now exhibited at the Museum of Practical Geology, Jermyn Street, wrote many years ago upon the artificial colouring of agates.*

The most common practice is to darken layers of agate and produce onyx. This is easily effected, as the *white* layers of the geodes and other chalcedonic masses being crystallized, or crystalline quartz, are impenetrable to liquids, whilst the other layers, being of a porous description, are permeable. The agates are therefore steeped in oil and sugar (formerly honey and water) and then gently heated, and in a few weeks' time the various layers, according to their porous nature, will have imbibed the saccharine matter, although no perceptible change will be noted. But on boiling the stones in sulphuric acid, the absorbed oil becomes carbonised, and turns them to a light or dark brown, or even black, in proportion to the quantity which has penetrated.

Pliny does not seem to have been aware of the necessity of the sulphuric acid, the secret of which, perhaps, the lapidaries of that day kept back, as he says : † "All precious stones in general are improved in brilliancy by being boiled in honey, Corsican honey more particularly, but acrid substances are in every respect injurious to them. As to the stones which are variegated, and to which new colours are imparted by the ingenuity of man, as they have no name in common they are called '*Physis* Nature,' *i.e.*, Works of Nature." He also says "'*Cochlides*,' probably petrified shells, are rather artificial productions than natural, found in Arabia, and which are boiled, it is said, in honey for seven days and seven nights," but does not mention the sulphuric acid, which is most important. To produce Carnelian or Sard, grey Chalcedony is steeped in pernitrate of iron, generally made by dissolving iron wire in nitric acid, and then subjecting it to heat, which turns it red. Brown Chalcedony, from either

---

* "*On the history of colouring Agates,*" V. Nöggerath, "*Die Kunst Onyxe zu färben. Karsten Archiv.*," xxii., 1848, *p.* 262
† *Lib.* xxxvii., *cap.*, 74.

India or Brazil, is reddened by being simply heated. The manner of producing dark-brown Sardonyx, or of staining the Onyx black, or deepening its colour, is by placing the stone into certain clear, colourless materials, such as oil or syrup, which will penetrate by the pores, and then boiling the stone in sulphuric acid, which chars or burns black *all organic substances* which the oil and syrup contain, namely—Carbon. The Orientals stained agates, but used honey instead of sugar or syrup, and made the fire take the place of the sulphuric acid. If the agates be made too dark, the colour can be withdrawn by the aid of nitric acid, and, if necessary, redarkened again. Agates and Chalcedony are even stained blue, and in Switzerland are sold often to the tourists as lapis-lazuli, from the Alps. The blue* is produced on exactly the same principles as the red, introducing peroxide of iron, in solution, but instead of heating them to turn red, precipitating the imbibed iron blue by soaking in a solution of ferrocyanide of potassium in the proportion of one scruple of ferrocyanide to an ounce of water, the *yellow* prussiate of potash, in water. Blue is also produced in other ways by protosulphide of iron, etc. A green colour, resembling Chrysoprase, produced lately through the present fashion of wearing that stone, is introduced into translucent Chalcedony by soaking it in a saturated solution of nitrate of nickel. Yellow is procured by hydrochloric acid acting upon the oxide of iron, which impregnates the stone by nature, but is a colour not often used, as the yellow obtained is generally pale, and will not show at all unless the Chalcedony is nearly transparent. A white stone, in fact, can be stained nearly any colour. A series of no less than twelve specimens, all of different decided colours and tints, are exhibited in the Museum of Practical Geology, Jermyn Street, Horse-shoe case), *cut from the same white stone.* Many so-called Turquoise sold by jewellers are Chalcedony artificially

---

* " *Science of Gems,*" *Dr. Archibald Billing*

treated. The Indians and Burmese have coloured stones artificially from the most remote periods, handing down by oral tradition their process from one generation to another, without their exact mode of procedure coming to light. Other stones, such as Fluor Spar are subjected at Oberstein to heat, the blue and voilet colours changing to purple, whilst a blue and a red would change to a reddish colour. Topazes, yellow are turned to pink, and dark Cairngorms to a light yellow—the heat acting upon the metals contained in them.

Although Oberstein has certainly been the means of bringing Agates, both in their natural and *unnatural* state, before the inhabitants of all the world, and deserves to be, therefore, congratulated, yet the author cannot refrain from expressing his great dissent from the manner in which the Agates and so-called "works of art" from Oberstein are exported. Lovely stones, both from South America and other localities, enter Oberstein in their natural state, to leave it either dyed of an unnatural colour, and robbed of all their beauty, or so worked into a vase or other object, and termed a "work of art," as to be really detrimental and certainly not complimentary to the taste of the nations to which it is exported. As a rule no attempt is made at proportion, the shape if a vase being generally under-sized or stunted, offensive to the eye, or, if correct in height, then, usually spoilt by suddenly bulging out in the centre, destroying all "lines" in its shape, or with a mouth so large that it is bigger at the top than the base, and might as well, if considered as a "work of art," be turned upside down.

Even if these imperfections are avoided, the object is generally spoilt by an objectionable unnatural colour of a deep red or blue, totally hiding the natural beautiful grey or fawn-coloured Chalcedony, perfect in Nature's own tints. The Agate bowls, for which Oberstein is so noted, are treated in the same manner, their colour either being deepened, and consequently losing all the charming softness which is one of their great characteristics of beauty, or they are stained, generally rendering the marvellous disposition of their zones or layers indistinct and muddy. It is not only by dyeing Agates, and making so-called " works of art," that the " Obersteiners' " work is to be deprecated, but because every mineral they procure and touch they treat in the same manner. Even the exquisite soft-toned Crocidolite, with its golden-yellow chatoyant colour (what stone could be more lovely in its natural state ?), has not escaped them, and it is painful to see the quantities of caskets, knick-knacks, umbrella and stick handles, in Europe, of this material dyed *blue, green,* and even *red,* destroying all their natural beauty and exquisite sheen, as well as softness of colour.

Agates, jasper balls, and other objects, are cut, stained blue, and sold in the Alps, at Homburg, and other watering-places, as *Lapis-Lazuli.* Thousands of their vulgar dyed " bulls-eyes," as Professor Ruskin calls them (Onyx black and white in strata or in concentric zones) are exported all over the world, and largely to Lake Superior, where they are unfortunately sold on the spot to the " 'cute Yankee " as local specimens, as well as to every part of America ; whilst in our own country, Brighton, Scarborough, and other well-known watering-places, are full of the fictitious coloured pebbles Even India, the cradle of all that is lovely in Agates, is inundated, as well as Burmah, with the dyed onyx-beads now manufactured at Oberstein, which are re-exported into England as Indian stones. It may naturally be answered, of course, that it is a matter of business, and that the " Obersteiners " only supply the demand made upon them. Granted ; but it must not be forgotten that they, by primarily exporting everything that was unnatural in the shape of colour and design, created that demand from the thoughtless and uninitiated, and are, in fact, still spreading a taste among the general inhabitants of Europe and, more particularly, of America, founded upon principles the very antithesis of all that is fundamentally correct in Nature and Art. The exportation of these artificially coloured objects is the more to be deprecated because it is unnecessary.

Professor Ruskin has had naturally always a horror of vulgar dyed specimens, and for some time the late Earl of Derby would not purchase *stained* specimens, but only added mineralogical examples—specimens in their perfectly natural colours—to the collection.

I feel assured that if the Germans would simply cut and polish their Agates, without staining and "boiling" them, they would find a readier market for them, particularly in England. No refined, educated lady or gentleman could admire a gaudy, artificially-coloured Agate bowl. As that *piquante* writer, "Vera Tzaritsin," in *The World*, remarked, on hearing a bowl she was holding was coloured: "Once I hear this, all the charm of colouring vanishes, and I reserve my admiration for an undyed bowl, which is, indeed, far more lovely with its exquisitely soft tones of fawn, white, and gold,"—and she only echoed the opinion of the educated classes.

There are exceptions to every rule, and I believe that the principal firm in Oberstein, represented by that well-known urbane agent, Herr Max Mayer, has lately produced objects of a much higher class, vases and bowls in their natural colours, and I feel assured in the long run, not only to his own, but to his manufacturers' great credit and benefit.

Some French firms have lately started mills on the Marne, and are turning out very fine works of art in Agate, Jasper, and other true *pietra dura*, with all that delicate gracefulness of form, perfect high finish, and taste for which the nation is so justly and universally celebrated.

AGATE.

'Αχάτης from the River Achates in Sicily, as described by Pliny, " *Reperta primum in Sicilia iuxta flumen eiusdem nominis.*"* It is also suggested that the name was derived from Accho, heated sand, or Akka, the western section of the plain of Esdraelon in Issachar. It was called Issachar, the eighth stone in the breast-plate of the High Priest. In Hebrew it is *Shebo*, and is, with all its varieties, called by the Italians *pietra dura*.

It is generally divided into three sections, banded, cloudy, and those specimens coloured by visible impurities. The banded varieties are very beautiful, occurring in delicate parallel lines ranging from pure white to black as well as occurring in different colours of brown, grey, fawn, and sometimes blue. One variety is called the Eye Agate, consisting of concentric circles, † *Leucopthalmus* of Pliny when consisting of one or two eyes, and the *Triopthalmus* ‡ when there are three. The concentric layers of these differ in porosity and are often artificially coloured. The white cloudy variety is the *Leucachates* (from λευκός = white) of Pliny ; the Wax Agate (cera = wax) is his *Cerachates*, § a name that was probably applied also to ordinary wax-coloured chalcedony ; the reddish Carnelian Agate is his *Sardachates*, and included the banded varieties ; a true light coloured Agate with red spots of Jasper was his *Hemachates*, ‖ (from αἷμα = blood) ; whilst his *Iaspachates* ¶ was an agate in which bluish or greenish shades predominated. Fortification and ruin agate are varieties with light to dark brown shades exhibiting when polished markings like fortifications and ruins. The varieties whose colours depend upon visible impurities are the Moss Agate and Mocha Stone, the *Dendrachates* of Pliny, from

---

* *Lib.* xxxvii., *chap.* 54.   † *Leucopthalmus, Lib.* xxxvii., 62.
‡ *Triopthalmus, Lib.* xxxvii., 71.   § *Cerachates, Lib.* xxxvii., 54.
Hemachates, Lib. xxxvii., 54.   ¶ *Iaspachates, Lib.* xxxvii., 54.

δένδρον — a tree—Chalcedony filled with brown and black moss-like forms desseminated through the mass.

Hardness 7·0. Specific gravity 2·58 to 2·66. Fracture conchoidal to uneven, and splintery.

Composition : Silicon with oxygen, with traces of metallic oxides.

An analysis by Redtenbacher of a Hungarian Chalcedonic Agate yielded :

Silica	98·81
Ferric Oxide ...	0·53
Carbonate of Lime	0·62
	99·96

No. 6. Agate.

No. 192. AGATE. Amygdaloidal, banded in parallel strata of white, bluish-grey, and black, called *Riband-Agate*. Section cut from a long amygdaloidal flattened Agate, assuming the shape of the cavity in which it was formed. The rough exterior exhibited top and bottom. Colours natural, not artificially treated. Transparent by transmitted light. The structure exhibiting perfect and distinct parallel layers or bands. Polished both sides. 9¾ by 4½ ins.

Coll. Arthur Wells.　　　　　Uruguay, South America.

A beautiful and very characteristic specimen of the Riband-Agate.

No. 193. AGATE. Banded or Riband-Agate, similar to No. 192. Strata in horizontal bands, white, bluish-grey, and black alternately. Translucent. 5½ by 4 ins.

Coll. Arthur Wells.　　　　　Uruguay, South America.

No. 194 GREY CHALCEDONY section with crypto-crystalline Silica divided by parallel bands of thin distinct white lines and layers of Chalcedony, disturbed probably through the intrusion of cross, irregular bands of ferruginous Chlorite —*Delessite*. Polished on one side. 6¼ by 3½ ins.

Cut at Oberstein on the Nahe, but from Brazil.

A remarkable and instructive specimen of great interest. The banded portion of the Agate is milky and bluish-white, nearly opaque on one-half of the stone, but is clear and translucent, of a bluish tinge, on the other, and is embedded *between* the grey translucent Chalcedony which forms the bottom of the Agate adjoining the exterior, which is covered with Chlorite, and the crypto-crystal-line Silica with small crystals of Quartz, which indicate the *interior* of the Agate. The white lines and the bands on the right are convex and regular until interrupted by a dark greenish intrusion of Chlorite, to the left of which the same white regular lines and bands appear but a full eighth of an inch *lower*, when they are again interrupted by a cross thick band of Chlorite which has squeezed and *raised* the three upper white bands *higher* and eliminated the two lower altogether. These three upper white lines become very small and are just visible, *lowered again*, separated by a thin cross vein of the Chlorite ; the whole five lines and bands after interruption by one other cross band continue in regular parallel lines to the end of the stone. This peculiar interruption of the zones, which may be compared to the "slip" or "fault" often seen in geological strata, is most probably due to the cross Chlorite bands intruding with force from the exterior through the Silica when not quite solidified, which would account for the *upheaval* of the lines, whilst perhaps the expulsion of the same Chlorite on arrival at the interior, by the force of combined gas, would cause the lowering or depression, as indicated by the second division of the lines and bands from the right.

The pyramids with their apices of the Quartz crystals are plainly discernible in the white Silica pointing inwards to the interior of the stone, which was cavernous, and surrounded by small crystals of Quartz.

No. 195 GREY CHALCEDONY. Section with white crypto-crystalline Silica divided by parallel bands of thin distinct white lines and layers of Chalcedony, disturbed probably through the intrusion of cross, irregular bands of ferruginous Chlorite, *Delessite.* Polished on one side. 6½ by 3½ ins.

Cut at Oberstein, but from Minas Geraes District, Brazil.

The same remarks apply generally to this Section as to the former (No. 194).

No. 196. CHALCEDONIC AGATE. Banded section in white, bluish-grey, and black parallel strata, tinged red with ferric oxide in the centre. Contains onyx-like circles of white and black concentric lines ; exhibits well the deposition of parallel bands, and exterior of stone. An analysis of a Chalcedonic Agate made by Redtenbacher yielded—

Silica	...	...	...	98·81
Ferric Oxide	...	...	...	0·53
Carbonate of Lime	...	...	...	0·62
				99·96

Well polished.   11½ by 5 ins.   Uruguay, South America.

No. 197. CHALCEDONIC AGATE. Section from the same stone as the preceding, banded white, grey, and black, with a little red in the centre. Presents an onyx appearance in concentric circles. One edge nearly straight, the other jagged. 11½ by 5¼ ins.

Uruguay, South America.

No. 198. CHALCEDONIC AGATE. Banded, section cut from same stone, also exhibiting onyx-like circles; thinner specimen. 11¼ by 4¾ ins.

Uruguay, South America.

No. 199. CHALCEDONIC AGATE. Section of white, blue, and grey Chalcedony, with flammeate red centre, probably caused by oxide of iron, encircling crypto-crystalline Silica. Translucent. Polished on one side. 7 by 3½ ins.

Uruguay, South America.

No. 200. CHALCEDONIC AGATE. Section of white, blue, and grey Chalcedony with flammeate red centre. Natural colour not artificially deepened. Similar to No. 199. Translucent. Polished on one side. 7 by 3½ ins.

Uruguay, South America.

No. 201. CHALCEDONIC AGATE. Magnificent amygdaloidal specimen in layers of white, blue, and grey, partly stained red with oxide of iron. Cut from a large, irregular oval-shaped stone, which depicts the shape of the cavity in which it was formed. Near the centre is a cavity encrusted with minute crystalline Quartz. A fine and important specimen. Translucent. Polished on one side. 15½ by 6 ins., nearly ½ in. thick.

Uruguay, South America.

No. 202. CHALCEDONIC AGATE. Magnificent amygdaloidal specimen in layers of a white, blue, and grey, partly stained red with oxide of iron. Cut from a large, irregular oval-shaped stone, which depicts the shape of the cavity in which it was formed. Near the centre is a cavity encrusted with minute crystalline Quartz. A fine and important example. Translucent. Polished on one side. 15½ by 6½ ins. nearly 3 ins. thick.

Uruguay, South America.

No. 203. CHALCEDONIC AGATE SECTION. Very fine reddish-brown, with small fortification. Opaque. Centre of Hornstone, surrounded by concentric layers, one of two nodules ; the other is of semi-crystalline Silica surrounded by Hornstone and concentric layers of Carnelian with Chalcedony. The outer edge has patches of flammeate Carnelian. Cut from an irregular distorted oval Agate. Translucent by transmitted light. Polished one side. 13 by 6 ins.

Uruguay, South America.

No. 204. AGATE SECTION, cut from a long, irregular amygdaloid, consisting of a white centre of concentric layers of white and grey Chalcedony, surrounded by black and white distinct strata. One side of edge irregular, the other rounded. Long and very interesting section. Well polished. 13¼ by 4 ins.

Uruguay, South America.

No. 205. AGATE SECTION, amygdaloidal, similar to No. 204. The centre consists, however, of nearly opaque Chalcedony, approaching to Hornstone, encircling a small cavity of crystalline Quartz. A thin long section. Well polished. 13¼ by 4 ins.

Uruguay, South America.

No. 206. AGATE SECTION cut from a long, irregular amygdaloid consisting of a white centre of concentric layers of white and grey Chalcedony, surrounded by black and white distinct strata of Agate. One side of edge irregular, the other rounded. A long and very interesting section. Well polished. 12 by 4 ins.

Uruguay, South America.

No. 207. CHALCEDONIC AGATE, amygdaloidal. Thick section cut from a long, flattened, irregular-shaped stone. The centre of white and grey Chalcedony surrounded by reddish-brown Carnelian or Sard. Exterior, or rough outer edge, exhibited, on one side of which are fortification lines. The Carnelian or Sard character is readily observed by transmitted light. Well polished. 11 by 2¾ ins.

Uruguay, South America.

No. 208. CHALCEDONIC AGATE. Magnificent section, with a chalcedonic centre surrounded by grey and bluish concentric layers, encircled by a thin band of reddish-brown Agate, the border of a mottled appearance of dark dendritic reddish-brown Agate, probably due to oxide of iron, with fortification bands. Section cut from a large, irregular stone. Translucent. Cut and polished both sides. 12 by 8 ins.

Uruguay, South America.

One of the finest and most important specimens in the collection, as well as one of the largest and broadest known.

No. 209. CHALCEDONIC AGATE. Very fine section, with small nodules of Chalcedony in the centre surrounded by grey and bluish concentric layers, encircled by a thin reddish band of Agate with a mottled border of dark dendritic Agate, due to metallic oxide, with fortification bands. Section cut from a large, irregular stone (similar to No. 208). Translucent. Cut and polished on both sides. 13¼ by 8 ins.

Uruguay, South America.

No. 210. CHALCEDONIC AGATE. Section of grey and bluish concentric layers, encircled by a band of mammilated Chalcedony and a thin pale reddish layer of Agate, with a mottled border of dark dendritic Agate. Section similar to No. 208 but smaller. Well polished on both sides. 11 by 6 ins.

Uruguay, South America.

No. 211. CHALCEDONIC AGATE. Smaller section of the same Agate as the three previous. A centre of Chalcedony with very deep black mottled border, giving it a dappled appearance. Cut and polished on both sides. 9⅓ by 4½ ins.

Uruguay, South America.

No. 212. CHALCEDONIC AGATE. Amygdaloidal. Long section of pale reddish-brown and white Chalcedony in fine concentric layers, which are very distinct. At one end tends towards Sard. Exterior of stone visible. Translucent. Polished on one side. 12 by 3¼ ins.

Uruguay, South America.

No. 213. CHALCEDONIC AGATE. Amygdaloidal. Another section of pale reddish brown and white Chalcedony in fine distinct concentric layers, cut from the same stone as No. 212. Translucent. Polished on one side. 12 by 3¼ ins.

Uruguay, South America.

A most characteristic specimen.

No. 214. CHALCEDONIC AGATE. A long section of pale reddish brown and white Chalcedony in fine distinct concentric layers similar to Nos. 212 and 213. Translucent. Polished on one side. 12 by 3¼ ins.

Uruguay, South America.

A very narrow amygdaloidal example.

No. 215. CHALCEDONIC AGATE. Another similar section. Long pale slice of reddish brown and white Chalcedony in fine distinct concentric layers, with a cavity in the centre of minute crystals of Silica. Slightly wider. Polished on one side. 10¼ by 3⅜ ins.

Uruguay, South America.

No. 216. AGATE SECTION. An abnormal and magnificent specimen, consisting of a cavity in the centre of crystalline Quartz, surrounded by beautiful filiform Silica and a wide band of bluish Chalcedony with a white line of fortification, encircled by a large layer of crypto-crystalline Silica, in which is another cavity of crystalline Quartz, the whole surrounded further by a lovely zone of beautiful violet semi-crystalline Amethyst paling off to white, encircled by an irregular band of minute zones of white Chalcedony, on one side of which there are "channels of egress" with "inlets of infiltration," the outer edge being of a Chalcedony looking like Carnelian or Sard. Most lovely specimen by transmitted light and of a striking pattern. One of the finest sections known. Cut from an exceptionally solid Agate. Polished on both sides and exhibiting outer edge. 15 by 9½ ins.

Rio Grande do Sul, Brazil.

No. 217. AGATE SECTION, consisting of a centre of greyish blue Chalcedony surrounded by a white fortification edge and wide band of beautiful crypto-crystalline Quartz, opaline by transmitted light, encircled by a lovely layer of semi-crystalline Amethyst of a bluish-purple paling off to white towards an outer rim of irregular Chalcedony. By transmitted light a most lovely specimen, the crypto-crystalline Silica and the Amethyst being particularly brilliant and translucent, whilst the Chalcedony looks like Carnelian or Sard. Polished on both sides. 15 by 9 ins.

Rio Grande do Sul, Brazil.

No. 218. AGATE SECTION, consisting of a centre of semi-crystalline Silica surrounded by band of bluish Chalcedony and a large beautiful layer of opaline crypto-crystalline Silica, with another band of lovely bluish purple semi-crystalline Amethyst, the prisms and pyramids of the crystals distinctly discernible, the whole encircled by minute concentric zones of bluish-white and brown Chalcedony. Similar to Nos. 216 and 217, but somewhat thinner and perhaps more beautiful by transmitted light. Polished on both sides and exhibiting outer edge. 15 by 9½ ins.

Rio Grande do Sul, Brazil.

No. 219. CHALCEDONIC AGATE. Irregular oval. Centre of bluish and light reddish-brown Agate, surrounded by concentric layers of white and blue Chalcedony with a large zone about an inch thick of blue Chalcedony encircled by grey layers, with an outside of pure opaque white Silica. Exhibits gaseous channels. By transmitted light has the appearance of Sard. Polished on one side. 11 by 7½ ins.

Uruguay, South America.

A most beautiful and interesting section. One of the gems of the collection.

No. 220. CHALCEDONIC AND CARNELIAN AGATE. A centre of Hornstone and white fortification Agate, surrounded by concentric layers of blue and grey Chalcedony, intruding into which are onyx-like projections of Carnelian and concentric Chalcedony, forming irregular nodules. Sard colour by transmitted light. Polished on one side. 10⅜ by 5½ ins. by ½ in. thick.

Uruguay, South America.

A most instructive specimen, exhibiting in its structure great interruption during formation.

No. 221. AGATE SECTION. Round but distorted. A cavity at one side is of pure white Chalcedony, surrounded by amorphous Silica, which is encircled by concentric zones of grey and white Chalcedony, with a darker border of the same. Waxy colour. Cut from a fairly round stone. Shows rough exterior. Polished both sides. 10 by 8 ins.

Minas Geraes, Brazil.

A very fine specimen, exhibiting, with marked distinctness, the separate depositions of the layers.

No. 222. AGATE SECTION. Round but distorted. Cut from the same specimen as No. 221. The white centre of Chalcedony is, however, more developed, surrounded by amorphous Silica and concentric zones of grey and white Chalcedony. Waxy colour. Exhibiting exterior. Polished both sides. 10 by 8 ins.

Minas Geraes, Brazil.

No. 223. AGATE. Consisting of an interior of beautiful crypto-crystalline Silica of an opaline tinge, with a border of Quartz crystals, coloured green, most probably artificially produced, with the apices of the pyramids pointing inwards, surrounded by bands of milky-white Chalcedony encircled by a deep band of grey Chalcedony. Very beautiful by transmitted light. Translucent. Outside waxy, like the *Cerachates* of Pliny. Irregular shape. Beautifully polished. 9 by 6¾ ins.

Minas Geraes, Brazil.

No. 224. Section of a circular ONYX AGATE, consisting of a centre of yellowish Chalcedony with a small core surrounded by countless *most minute* concentric layers of grey and bluish-white Chalcedony to a diameter of two inches, which is then encircled by a further series of minute white zones alternate with bands of grey Chalcedony to a diameter of three inches, forming a circular Onyx which is embedded in an irregular border, running about an inch, of crypto-crystalline Silica of an opaline tinge by transmitted light, with a cloudy, milky-white edge. The apices of the crystals point outwards and are distinctly visible, running into a clouded band of milky-white Chalcedony, which is spotty by transmitted light. The whole encircled by white semi-crystalline Silica, exhibiting in places the hexagonal form of the crystals. 8 by 6½ ins.

Minas Geraes, Brazil.

The centre, by transmitted light, is of a reddish-brown carnelian, and illustrates the minute deposition of zones. One of the finest and most beautiful specimens in the "Derby" Collection and of interest as being the first specimen purchased by the late Earl Derby, and was in fact the foundation of the whole series.

No. 225. AGATE, conical, consisting of a brown surface with irregular circles and bands of white and bluish-white Chalcedony, irregularly dispersed, some wavy, others zig-zag, and in irregular concentric layers. Both sides polished, showing the same markings. The base, which is unpolished, proves the Agate to be of a Sardonyx nature, consisting of a whity-brown centre surrounded by alternate red and white zones, the outside being Sard. Height 6¼ ins., width 8 ins., depth 3¼ ins.

Uruguay, South America.

No. 226. AGATE, thin large amygdaloidal section. The centre is composed of transparent crypto-crystalline Silica of an opaline tinge by transmitted light embedded in white strata, forming a fortification which is surrounded by extremely minute concentric zones of white, bluish-grey Chalcedony, showing clearly every minute separate deposition, and further encircled by broader bands composed of minute strata of brownish Carnelian Agate tinged a deep red by iron oxide at the edges. Transparent to translucent. Exhibits mammillated structure by transmitted light. Polished on both sides, and only the eighth of an inch in thickness. 15½ by 7¼ by ⅛ in. in thickness.

District of Minas Geraes, Brazil.

This abnormal section deserves particular attention through its unusual size and beauty, as well as representing the great perfection to which the cutting of Agates has attained in France. Although only the eighth of an inch in thickness, and fifteen and a half inches long, it has been cut out of the *centre* of a large stone with a specially constructed thin wheel, highly impregnated with Diamond dust, driven by enormous hydraulic power. Then, most highly polished on both sides, and all without a single crack or flaw. Such a perfection of cutting and polishing as exhibited in this Agate, as well as in Nos. 228, 229 and 230, has only recently been developed in France leaving Oberstein far behind, whose powers now are devoted, unfortunately, to the production of the cheaper and artificially coloured specimens.

No. 227. Section of two AGATES, irregularly round, joined together at one side. In the interior of each are the remains of a Coral (carbonate of lime), which has become silicified, further encircled by reddish-brown Carnelian and pale-blue Chalcedony. In the larger section are nodules of Carnelian (Sard) the centre of which contains the remains of the Coral, which is surrounded by transparent Chalcedony. The smaller section, also, has the white Coral in the centre, surrounded by well-marked Chalcedony and Carnelian. The outside edge of both sections is exhibited except where joined. 9¾ by 7⅝ ins. Polished.

Minas Geraes, Brazil.

No. 228. CHALCEDONIC AGATE section, nearly round, extremely thin, consisting of a centre of transparent crypto-crystalline Silica with an opaline tinge, surrounded by *most minute* concentric zones or layers of pale bluish and white Chalcedony, in marvellous regularity and perfectly distinct. Very translucent ; nearly transparent. Beautifully polished on both sides. 10¼ by 9 ins., but only 1-16th of an inch in thickness.

Brazil.

The thinnest and one of the most beautiful sections in the collection, illustrating, as No. 226, the marvellous minuteness of the deposition of layers as well as the great perfection which the art of cutting and polishing has attained in France. The section is only about the *one-sixteenth part of an inch* in thickness, cut from the centre of a stone nearly round, and polished to the highest degree on both sides.

No. 229. AGATE section, irregular shape, exhibiting a deep black fortification centre, in the middle of which are three thin white tri-radiate lines surrounded by transparent and grey concentric zones of Chalcedony, with a broad band of deep black, encircled by a waxy-looking stratum of minute zones of Agate running into a border of reddish-brown Sard. Exhibits shape of stone. Very translucent. Very thin section. Exceptionally well polished on both sides. 9⅛ by 7 ins.

Uruguay, South America.

The colour in this section is quite natural, not having been deepened by artificial means. This the author is assured of by the lapidary

who received it in Paris direct from Brazil, and who cut and
polished it personally on the Marne, near Paris, where artificial
means for deepening or dyeing Agates are at present unknown,
or at least not practised. Had the section been cut in Germany,
most authorities would have pronounced the colour as artificial.
Attention is directed to the beautiful polish of this Agate on both
sides.

No. 230. AGATE section, irregular shape, similar to No. 229,
being cut from the same stone. The black centre is, however,
smaller, whilst the tri-radiate white lines are more freely
developed. The black fortification is encircled by transparent
layers of grey and reddish Chalcedony, approaching to Carnelian at the edges. Transparent. Cut very thin and beautifully polished on both sides. Exhibiting by transmitted light
formation of stone. 6¼ by 6¼ ins.

Uruguay, South America.

The remarks following No. 229 naturally apply to this section. It
was cut and polished by the same French firm.

No. 231. Section of AGATE, irregular shape, cut from the
same stone as Nos. 229 and 230, with a fortification in
the centre of black, surrounded by transparent and grey
concentric zones of Chalcedony. Similar to Nos. 228 and
229. Extremely well polished on both sides. Translucent.
6 by 6 ins.

Uruguay, South America.

No. 232. AGATE, fine thin section of beautiful transparent
crypto-crystalline Silica of a greenish tinge, surrounding the
centre of a map-like bluish-grey fortification Chalcedony tipped
with red. At the base are milky cloud-like and reddish
bands of irregular Chalcedony. The stone was irregular,
nearly flat at the base, the original shape being shown by the
exterior layer, which is red. Translucent in the centre.
The Silica transparent. Colour quite natural. Most highly
polished. 6 by 5 ins.

Brazil.

A lovely section, beautifully polished on both sides in France.
Presented by the present Earl of Derby.

No. 233. AGATE, fine thin section of beautiful transparent crypto-crystalline Silica of a pale green tint, surrounding a map-like bluish-grey Chalcedony, tipped with red. At the base are cloud-like bands of Chalcedony, similar to No. 232. The junction of the semi-crystalline Silica is observable. Base very red. Cut and well polished on both sides in France. Translucent. $5\frac{1}{2}$ by $4\frac{1}{2}$ ins.

Minas Geraes, Brazil.

Presented by the present Earl of Derby.

No. 234. AGATE section. Highly coloured flamboyant red. The centre of amorphous transparent Silica of an opaline tinge by transmitted light, surrounded by concentric strata of Chalcedony, which is further encircled by a red mossiform-looking border of opaque Agate running into Jasper. A thin section. Beautifully polished on both sides in France. $5\frac{3}{8}$ by $3\frac{1}{2}$ ins.

Minas Geraes, Brazil.

Presented by the present Earl of Derby.

No. 235. AGATE section, similar to No. 234, of highly coloured flamboyant red. The centre of amorphous transparent Silica of an opaline tinge by transmitted light, surrounded by concentric layers of Chalcedony and red mossiform-looking border of opaque Agate running into Jasper. Polished on both sides. $5\frac{1}{2}$ by $3\frac{1}{2}$ ins.

Minas Geraes, Brazil.

Presented by the present Earl of Derby.

No. 236. Dark AGATE section, with irregular brown map-like patches, and a grey-whitish patch which appears to be, upon close examination, of a coralline structure. A thin section, exhibiting exterior of stone. Opaque. Well polished on both sides in France. $5\frac{3}{4}$ by $5\frac{1}{2}$ ins.

Minas Geraes, Brazil.

Presented by the present Earl of Derby.

No. 237. Dark AGATE section, similar to No. 236, with map-like brown patches and a whitish patch which appears to be a silicified Coral. A thin section, exhibiting exterior of stone. Opaque. Well polished on both sides in France. 5¾ by 5½ ins.

Minas Geraes, Brazil.

Presented by the present Earl of Derby.

No. 238. AGATE. Half of a fine circular stone exhibiting a centre of grey surrounded by concentric layers of pale blue white Chalcedony and brown Carnelian. A most regular and extremely well-marked specimen, the layers being most distinct, the light Chalcedony being particularly striking, and regular with the reddish Carnelian. Shows complete exterior of one half of stone. Face polished. Opaque. 6¾ by 5¾ ins.

Minas Geraes, Brazil.

No. 239. AGATE BLOCK, cut from a large specimen. The sides exhibit filiform amorphous amethystine Silica, surrounded by strata of Chalcedony deeply tinged with red oxide of iron. The top flat side presents a mottled appearance of dark nodules or circles surrounded by deep vivid blood-red Jasper, with Amethyst occasionally intervening. Polished on all sides but base, which exhibits outside crust. 6 by 4 ins.

Minas Geraes, Brazil.

A remarkable and deeply interesting stone.

No. 240. AGATE of a vescicular character composed of pale blue Chalcedony, with white Agate on the border. The exterior is cavernous and vescicular, and exhibits crystalline Quartz in a cavity. 5¼ by 3 ins.

Minas Geraes, Brazil.

No. 241. AGATE pebble. One half with face cut and polished, exhibiting grey and dark bluish Chalcedony in alternate layers. 3¾ by 2½ ins.

Uruguay, South America.

No. 242. AGATE cut from a Quartzite pebble. The centre is of amorphous Quartz, slightly violet, with small irregular bands of Agate surrounded by a broad layer of semi-crystalline Silica, with an outer edge of pinkish Agate. Polished. 5 by 3½ ins.

Aussig, Bohemia.

No. 243. AGATE, thin transparent section, with a centre of amorphous Silica, partly opaline by transmitted light, surrounded by a band of white Chalcedony, and a deep layer of violet-red Agate with a bluish rim. Translucent. Polished on one side. 4 by 3 ins.

Schemnitz, Hungary.

No. 244. AGATE SECTION, small. A crystalline cavity surrounded by a well-marked stratum of Hornstone and white Agate, with alternate bands of Chalcedony, parallel at base, but forming irregular fortifications at the top. The layers are very distinct. Section cut from the centre of a pebble. Translucent. Polished on one side. 4¼ by 2½ ins.

Coll. Dr. Birdwood.   Banda, India.

No. 245. AGATE SECTION, cut from the interior of a " Pebble," consisting of a crypto-crystalline centre of Silica, surrounded by bands of grey Hornstone and blue Chalcedony on a lake-like base. Polished. Translucent, looking like Sard. 4½ by 3½ ins.

Province of Minas Geraes, Brazil.

A very interesting specimen. The Agate has, during its formation, rested upon a flat surface, or has solidified first at the bottom instead of all round, as the layers forming the base are flat and parallel, not curved, whilst the remainder are irregularly concentric.

No. 246. AGATE SECTION. Irregular, opaque, white map-like formation in the centre, surrounded by amorphous Silica resting upon parallel lake-like strata, with outside edge of Carnelian. Translucent. Polished. 4¼ by 2¾ ins.

Minas Geraes, Brazil.

This Agate lays also upon a horizontal base, showing it must have solidified there first, instead of all round—the strata of Chalcedony and Carnelian being parallel, or " lake-like," instead of curved.

II

No. 247. AGATE SECTION. The centre of most minute layers of dark brown Chalcedony in an oblong circle, with Agate of a reddish tinge surrounded by a broad band of blue-white Chalcedony, which is encircled at the top with amorphous transparent Silica, with an opaline tint. Polished on one side. $5\frac{1}{2}$ by $3\frac{3}{4}$ ins.

Minas Geraes, Brazil.

No. 248. AGATE, section of dull milky-white, and pale bluish Chalcedony, with outer rim of concentric strata of Carnelian (Sard). Polished on one side. $4\frac{1}{4}$ by $2\frac{7}{8}$ ins.

Oberstein on the Nahe.

No. 249. Section of AGATE with a centre of amorphous Silica, surrounded by minute concentric strata of Chalcedony and Carnelian. Reddish waxy brown by transmitted light. Polished on both sides. $4\frac{1}{2}$ by 4 ins.

Minas Geraes, Brazil.

No. 250. Thick section of AGATE. The centre is red fortification Agate, a band of semi-crystalline Silica, and a border of brown Agate encircled by amorphous Silica and pale blue Chalcedony. A striking and pretty specimen. Polished on one side. $4\frac{1}{2}$ by $3\frac{1}{2}$ ins.

Minas Geraes, Brazil.

No. 251. AGATE. Exhibiting the half of a pebble with band of brown and layer of milky-white Chalcedony. Polished on surface. 4 by $3\frac{1}{2}$ ins.

Minas Geraes, Brazil.

No. 252. Irregular inlaid AGATE of Chalcedony. Agate associated with brown Carnelian upon a base of rough blue Chalcedony. Polished on one side. $3\frac{1}{2}$ by $3\frac{1}{2}$ ins.

Minas Geraes, Brazil.

No. 253. Waxy-looking AGATE in bands of fortification Chalcedony and Carnelian, associated with crypto-crystalline Quartz. Polished one side. $4\frac{1}{4}$ by $3\frac{3}{8}$.

Galgen-Berg, Oberstein.

No. 254. Amethystine AGATE with semi-crystalline centre, surrounded by extremely minute layers of Chalcedony. Upon a base of pure Chalcedony with a vescicular surface. Polished on one side. $3\frac{1}{4}$ by 3 ins.

Galgen-Berg, Oberstein.

No. 255. Irregular semi-crystalline AGATE, dark and white, with border of Carnelian, exhibiting rough exterior, therefore the shape of the stone. 3 by $2\frac{3}{4}$ ins.

Minas Geraes, Brazil.

No. 256. Salmon-coloured AGATE, in circles and partly in fortifications stained with a metallic oxide, associated with black map-like markings and semi-crystalline Quartz. 4 by 3 ins.

Aussig, Bohemia.

No. 257. Irregular AGATE of a buff-brown centre, with a distinct layer of white Chalcedony with outside of dark Carnelian. One surface polished. $3\frac{1}{2}$ by $2\frac{3}{4}$ ins.

Galgen-Berg, Oberstein on the Nahe.

No. 258. Section of chalcedonic AGATE, consisting of minute regular concentric layers of white and pale blue Chalcedony, with a rim of transparent Agate. Translucent. Polished on both sides. $5\frac{1}{4}$ by $4\frac{1}{4}$ ins.

Meerut district, India.

No. 259. AGATE, irregular, of milky and blue Chalcedony, cut from half a nodule ; at one side is evidence of a silicified coral. 3 by $2\frac{1}{2}$ ins.

Brazil.

H 2

No. 260. AGATE, half of an irregular nodule, of white and blue Chalcedony, with white Hornstone on the edge. One side polished. 3¼ by 2¼ ins.

Aussig, Bohemia.

No. 261. AGATE, consisting principally of semi-crystalline Quartz, in the centre of which is an irregular circle of brown Agate, with milky white Chalcedony resting upon strata of Chalcedony and Carnelian. Polished. 3¾ by 3¾ ins.

Minas Geraes, Brazil.

A well-marked and striking agate.

No. 262. Fortification AGATE, irregular. Pale blue Chalcedony in layers, surrounded by a band of black Agate in pale blue Chalcedony. Exhibits exterior. 3 by 2¾ ins.

Oberstein, Germany.

No. 263. Section of inlaid AGATE, of amorphous white Silica, with brecciated red and white nodules. Brown and red. Translucent. 4 by 2 ins.

Aussig, Bohemia.

No. 264. AGATE, irregular, in semi-fortifications of pink and white Agate, in minute irregular concentric zones. Exhibits exterior. 3¼ by 2½ ins.

Aussig, Bohemia.

No. 265. AGATE, with a cavity of minute crystals of Quartz, surrounded by amorphous and semi-crystalline Silica. Spotted red in patches, with a thin red layer on the exterior. 2¾ by 2½ ins.

Province of Minas Geraes, Brazil.

No. 266. AGATE of white amorphous Silica, with concentric bands of creamy-white Chalcedony, with a bluish border on the outside of a vescicular character. One side polished. 3½ by 3 ins.

Aussig district, Bohemia.

No. 267. AGATE of a pink or salmon colour, with white circles in irregular map-like formation upon a matrix of semi-crystalline white Quartz. 3 by 2½ ins.

Aussig district, Bohemia.

No. 268. AGATE, half of a pebble. A peculiar clouded uncertain formation of a bluish-white in irregular patches, with an outside border of Carnelian or Sard. 3¼ by 2½ ins.

Minas Geraes, Brazil.

No. 269. AGATE. Section of pale grey Chalcedony with border of opaline, semi-crystallized Silica. Translucent. Polished. 4 by 3 ins.

Minas Geraes, Brazil.

No. 270. AGATE section, with a milky clouded crypto-crystalline centre, with a border of Carnelian resting on a flat base. Translucent. 3¼ by 2¼ ins.

Province of Minas Geraes, Brazil.

No. 271. Thin section of AGATE consisting of irregular minute concentric layers of Chalcedony surrounded by transparent white Agate. The minuteness of the deposition of the layers is shown by transmitted light. 3¼ by 2½ ins.

Neighbourhood of Reykjavik, Iceland.

Brought by Dr. Tayler, the well-known Greenland explorer.

No. 272. AGATE, brecciated, with milky-white Chalcedony. Translucent to nearly opaque. Polished on one side. 2½ by 2½ ins.

Kangertlular Fiord, Greenland.

Coll. Captain Pulleine.

No. 273. Fortification red jaspery-looking AGATE, in concentric layers, with white Cacholong. Very pretty. Distinct markings. Opaque. Well polished. 2¾ by 2⅛ ins.

Aussig, Bohemia.

No. 274. AGATE. Irregular in shape, with a broad white band of Hornstone, resting upon a flattened base of parallel bands of Agate, Chalcedony, and Carnelian. Translucent to opaque. Carnelian colour in parts by transmitted light. $3\frac{1}{2}$ by $2\frac{1}{8}$ ins.

Banda, India.

Coll. Arthur Wells.

No. 275. Portion of a violet AGATE, cut from a convex stone, in reddish and pale blue strata. Irregular. Polished on one side. $3\frac{1}{8}$ by $1\frac{1}{4}$ ins.

Minas Geraes, Brazil.

No. 276. CHALCEDONIC AGATE. Thin section. Grey with dark, irregular patches. Translucent. Polished on one side. $3\frac{3}{4}$ by $2\frac{3}{4}$ ins.

Banda, India.

Coll. Arthur Wells.

No. 277. AGATE. Rectangular of a sandy, yellowish, clouded colour. Brilliant and opaline by transmitted light. Bevelled edge. Polished all over. $3\frac{1}{4}$ by $2\frac{3}{8}$ ins.

Meerut, India.

No. 278. AGATE. White amygdaloidal, in semi-fortifications. Opaque. Polished on one side. 3 by $1\frac{3}{4}$ ins.

Aussig, Bohemia.

No. 279. CHALCEDONIC AGATE, of a transparent grey, embedding red Jasper, with crystalline Silica. Polished on one side. $3\frac{1}{2}$ by 2 ins.

Minas Geraes, Brazil.

No. 280. AGATE, oval, with a dark centre, surrounded by concentric lines of brown Chalcedony. By transmitted light is a red Carnelian. Waxy. Polished all over. $2\frac{3}{4}$ by 2 ins.

Banda, India.

No. 281. AGATE, oval, with a well-defined fortification of whitish and Chalcedonic strata in wax-like bands of Carnelian (Sard). Polished all over. $2\frac{1}{8}$ by 2 ins.

Bombay, India.

No. 282. AGATE, oval, consisting of beautiful inlaid greyish-yellow, dark brown, whitish, and semi-crystalline strata. One edge of translucent amorphous Silica. Polished all over. Translucent to opaque. $2\frac{1}{8}$ to $1\frac{5}{8}$ ins.

Banda, India.

No. 283. AGATE, octagonal, with irregular-shaped fortification of minute bands of white and brown Chalcedony. Polished all over. $2\frac{1}{8}$ by $1\frac{7}{8}$ ins.

Cambay, India.

No. 284. Thick portion of AGATE (pebble); Chalcedonic centre, with dark band of Agate and minute white zones. The colour deepened artificially. $3\frac{3}{8}$ by $1\frac{1}{8}$ ins.

Cut at Oberstein, but from Uruguay, South America.

No. 285. AGATE SECTION. The centre composed of beautiful fortification grey and white Agate, surrounded by minute concentric layers of pale bluish Chalcedony, with deep bands consisting of minute zones of white Agate encircled by brownish translucent Carnelian. A remarkably beautiful section in very distinct strata. Irregular ; the formation is broken by a "channel of egress." Polished on one side. 4 by $3\frac{3}{4}$ ins.

Minas Geraes, Brazil.

No. 286. BRECCIATED AGATE, section of. Cut from a "pebble" consisting of opaline crypto-crystalline Silica, surrounded by reddish Jasper, associated with a limestone on the border. A nodule entering into the Jasper is of a salmon colour in fortification. $4\frac{1}{2}$ by 4 ins.

A good example of inlaying.

Minas Geraes, Brazil.

No. 287. WHITE AGATE, oval specimen of, in beautifully small concentric strata, embedded in milky translucent Chalcedony, showing mammillated form by transmitted light. Cut and polished all over, with bevelled edge. Transparent to translucent. $4\frac{1}{4}$ by $3\frac{3}{8}$ ins.

Minas Geraes, Brazil.

No. 288. CHALCEDONIC FORTIFICATION AGATE, oval. Consisting of a centre of white concentric Agate, surrounded by bluish-grey and white Chalcedony forming a complete fortification—embedded in translucent bluish Chalcedony. The centre fortification exhibits one or two "inlets of infiltration" or "channels of egress." Polished on both sides.

Minas Geraes, Brazil.

A remarkably beautiful specimen, illustrating the regular deposition of strata or zones in a marked degree.

Presented by the present Earl of Derby.

No. 289. AGATE, puce coloured, in irregular fortification, a solid centre surrounded by semi-crystalline Quartz, encircled by white onyx, and minute pink layers of Agate, with Delessite on the exterior. Well polished. 3 by $2\frac{5}{8}$ ins.

Aussig, Bohemia.

No. 290. CHALCEDONIC AGATE, bluish-grey and white, with slightly marked fortification on a true base of waxy Chalcedony. Polished. $2\frac{3}{4}$ by $2\frac{3}{4}$ ins.

Aussig, Bohemia.

No. 291. FORTIFICATION AGATE, in beautiful creamy-white, triangular strata, embedded in grey Chalcedony and surrounded by Carnelian. Polished all over. A well-marked specimen. $2\frac{3}{8}$ by 2 ins.

Banda, India.

No. 292. DARK AGATE, colour deepened artificially, associated with white opaline, crypto-crystalline Silica, of which one crystal is very distinct, showing apex of pyramid. Polished one side. 3 by 2 ins.

Cut at Oberstein, but from Brazil.

No. 293. AGATE (partly Hornstone), of an irregular shape with a grey centre, surrounded by bands of lilac. Opaque. One side polished. 2¼ by 1¾ ins.

Meerut, India.

No. 294. AMYGDALOIDAL AGATE. Opaque, associated with translucent Chalcedony running into Carnelian. A good example of embedding. Well polished. 4⅛ by 2 ins.

Minas Geraes, Brazil.

No. 295. CHALCEDONIC AGATE in very distinct strata of a milky-white, embedded in Chalcedony. The layers exhibit an interruption or "fault." Opaque and also translucent. 2¾ by 1½ ins.

Near Reykjavik, Iceland.

No. 296 to 304 inclusive. AGATE, sections of, nine, cut from the same pebble, exhibiting by gradation the formation of an Agate from its centre to the exterior. The larger section, No. 296, which was the centre of the pebble, consists of a small fortification of milky Quartz, surrounded by opaline amorphous silica, with a straight base resting upon parallel cloudy strata of alternate white Agate and Carnelian, slightly dipping, which runs to the bottom of the Agate, the whole encircled by a band of brown Sard, well developed in layers at the top, but thinner towards the base ; translucent. Section No. 297, the next in size, exhibits the fortification with a more decided brown fortification in the centre of the Silica, which is opaline and semi-crystalline, resting, like No. 296, upon parallel strata of Carnelian and Agate, slightly dipped ; translucent. Section No. 298, which is smaller than the preceding, is more like No. 296, having lost the decided fortification in the centre, but developing a cavity, the parallel layers forming a larger base ; translucent. Section No. 299 is a still smaller section, showing the interior of the fortification which has assumed a cavity with small crystallized Quartz crystals. The Silica around is less crystalline, whilst the base is in straight parallel layers. No. 300, a still smaller section, getting nearer the exterior, has crystals of Quartz in the fortification, which has become elongated. The Silica is not so large, the border of Sard being more developed at the top, whilst the base is still in straight parallel layers. No. 301, a still smaller section, has crystals in fortification, not so well developed; the straight parallel base not so well defined excepting at the bottom, where they dip, the Sard band being of a good colour. No. 302, another smaller is a more translucent section, the fortification being less and the parallel bands getting naturally smaller. No. 303, the smallest section but one, has the Silica disappearing and the parallel base interrupted; the outside Sard is more developed and shows a good reddish-brown by transmitted light. No. 304, the last and smallest section, exhibits the rough exterior of the pebble, the cavity still showing, but the Silica has disappeared, whilst the parallel base is broken but still showing its formation. A good Sard colour. All the sections are translucent, and illustrate well the gradual development of the deposition or growth of one-half of the pebble. Two specimens are polished on both sides, the seven others on one. The smaller section is $2\frac{1}{4}$ by $1\frac{1}{4}$ ins. The larger section is $3\frac{1}{4}$ by $2\frac{1}{4}$ ins.

Minas Geraes district, Brazil.

A deeply interesting series of graduated sections of a specimen, which cannot fail to arrest the attention of all students of the formation and genesis of the Agate.

No. 305. FORTIFICATION AGATE, of white and bluish Chalcedony, in minute concentric layers. Spotted on one side with iron, but well marked. Partly opaque. Polished on one side. $2\frac{1}{4}$ by $2\frac{1}{4}$ ins.

Aussig, Bohemia.

No. 306. AGATE, consisting of a crypto-crystalline centre, surrounded by a wide band of white Agate forming a solid fortification; surrounded by a broad exterior layer of reddish-brown Carnelian. Well polished. $3\frac{1}{2}$ by $2\frac{3}{4}$ ins.

Minas Geraes, Brazil.

No. 307. Cross section of Stalactite of white AGATE. Transparent opaline silica surrounded by a band of Chalcedony. Nearly round. Well polished. $1\frac{7}{8}$ by $1\frac{6}{8}$ ins.

Minas Geraes, Brazil.

No. 308. Section of variegated coloured AGATE, consisting of fortifications of brown, red, and white. Embedded with Jasper of various tints. Very pretty. Cut for the top of a snuff-box. Polished all over; with bevelled edge. 3¼ by 2¾ ins.

Campsie Hills, Scotland.

No. 309. AGATE. Chalcedony in broad, lake-like white bands in parallel layers, associated with crypto-crystalline Silica. Polished. 2 by 1¼ ins.

Reykjavik, Iceland.
Coll. Dr. Tayler.

No. 310. AGATE of white Chalcedony, and brown Carnelian. Polished on one side. 1½ by 1¾ in.

Reykjavik, Iceland.
Coll. Dr. Tayler.

No. 311. EYE AGATE, *Leucopthalmus* of Pliny.* Section of irregular concentric bands of grey and bluish-white Chalcedony, zigzag and waving, associated with crypto-crystalline Silica. Opaline by transmitted light. The centre of the eyes possesses cores of carbonate of lime, probably the remains of a coral. Translucent. Well polished on both sides, showing complete outside crust—therefore the original shape of Agate. Irregular. 6½ by 6 ins.

Province of Minas Geraes, Brazil.

No 312. EYE AGATE, *Leucopthalmus* of Pliny.† Section similar to No. 311, cut from the same stone. Bands of different tinted Chalcedony in distorted zigzag form associated with crypto-crystalline Quartz, in which there is a cavity of crystallized Silica. Cores in the centre of the circles of carbonate of lime. Well polished on both sides. 6½ by 5¾ ins.

Minas Geraes, Brazil.

* *Lib.* xxxvii., 62.　　† *Lib.* xxxvii., 62.

No. 313. Section of EYE AGATE, probably the *Triopthalmus* of Pliny, xxxvii., 71, consisting of circular eyes, three
ranging together of bluish-grey translucent Chalcedony, the
centre of each circle consisting of cores of carbonate of lime
in very minute concentric layers surrounded by white and
milky bands of Chalcedony. Translucent. Polished on both
sides. A very beautiful specimen. $6\frac{1}{2}$ by $5\frac{1}{4}$ ins.

Minas Geraes, Brazil.

No. 314. EYE AGATE, probably the *Triopthalmus* * of
Pliny. Consisting of circular eyes, three ranging together of
bluish grey translucent Chalcedony. Section similar to No. 313,
but thinner and more translucent. The circles contain cores
of carbonate of lime, probably the remains of a coral. Transparent to translucent. Beautifully polished on both sides
$6\frac{1}{2}$ by 5 ins.

Minas Geraes, Brazil.

No. 315. EYE AGATE, *Triopthalmus* of Pliny. Irregular
concentric eyes of bluish grey Chalcedony, occurring three
together, the centres having cores of carbonate of lime. A
map-like zigzag formation of a milky colour exists between
the eyes, which is opaque by transmitted light, whilst the
eyes are translucent. Well polished on both sides. 6 by 5 ins.

Minas Geraes, Brazil.

No. 316. EYE AGATE, perhaps the *Leucopthalmus* of Pliny.
Section consisting of irregular circles of greyish Chalcedony, surrounded by broad distinct zones of milky and white
Agate, which is embedded in filiform transparent Silica. The
centres of the eyes, some of which occur double, are encircled
by a white exterior, probably of lime silicified. Translucent. Polished both sides. A beautiful section in every way.
$5\frac{1}{2}$ by $5\frac{1}{2}$ ins.

Minas Geraes, Brazil.

No. 317. EYE AGATE, *Leucopthalmus* of Pliny.+ With a
nodule of probably carbonate of lime in a cavity in the interior surrounded by a concentric circle of brown and bluish
Chalcedony spreading out to a zigzag formation, associated
with amorphous Silica. Translucent. Carnelian-looking by
transmitted light. Beautifully polished. 3 by 3 ins.

Minas Geraes, Brazil.

* *Lib.* xxxvii., 71. | *Lib.* xxxvii., 62.

No. 318. EYE AGATE. Half of a pebble consisting of eyes single and double, embedded in grey Chalcedony, surrounded by strata of white and brown Chalcedony. One side polished. 2½ by 1¾ ins.

Minas Geraes, Brazil.

No. 319. EYE AGATE, consisting of two circles joined together; the smaller with a *brown* centre surrounded by white, the larger with a *white* centre surrounded by brown, encircled by a band of white which joins that of the smaller, the whole embedded in brown Agate and white Chalcedony. The reverse side is peculiar, of quite another colour, being grey and bluish-white. Polished all over. 2⅔ by 1¾ ins.

Minas Geraes, Brazil.

No. 320. EYE AGATE sections, consisting of two semi-circles of greyish Chalcedony joined together with a centre of brown Carnelian, the outer part being similar. The other side fainter in colour, with no band. Polished on both sides. 1¾ by 1½ ins.

Minas Geraes, Brazil.

No. 321. EYE AGATE. Oval, of a whitish milky colour. Many eyes, with zigzag fortification of white Agate. Translucent. Polished all over. 2⅛ by 1½ ins.

Bandah, North-West Province of India.

No. 322. Section of white amygdaloidal AGATE, cut from a long flattened boulder on flat base. Consists of an irregular cavity, surrounded by most beautiful minute fringed fortification of white, encircled by irregular bands of white-bluish Chalcedony. Opaque to translucent. This is probably the *Leucachates* of Pliny, from λευκός=white. 11¾ by 3⅞ ins.

Cut at Oberstein, but from Brazil.

No. 323. WHITE AGATE, of alternate bands of white and blue Chalcedony, enclosing a crypto-crystalline centre, the whole resting upon a parallel base. The exterior shows that the specimen consists of a nodule of Chalcedony, surrounded by a thick crust of the same material. Opaque. Well polished. 3½ by 1¾ ins.

Uruguay, South America.

No. 324. WHITE AGATE, probably the *Leuchates* of Pliny. Convex specimen of irregular layers surrounded by white border, tinged on the exterior with iron oxide. Opaque and well polished, with thin black lines. $3\frac{1}{4}$ by 2 ins.

Aussig, Bohemia.

No. 325. WHITE AGATE. Rectangular, with bevelled corners. Chalcedony in broad bands. Cut and polished all over. $2\frac{3}{4}$ by $1\frac{3}{4}$ ins.

Bombay, India.

Coll. Captain Pullcine.

No. 326. WHITE AGATE, rectangular slab (*Leuchates* of Pliny), in broad milky bands, with bands of bluish Chalcedony. Translucent. Cut and polished at Oberstein. 4 by $2\frac{3}{8}$ ins.

Banda, India.

No. 327. AGATE SECTION, consisting of fortification of amorphous Silica, surrounded by white and grey zones or bands of Chalcedony embedded in a waxy Agate, probably the *Cerachates* * of Pliny. Thick. Polished on both sides. Translucent. 5 by $4\frac{1}{4}$ ins.

Minas Geraes, Brazil.

No. 328. AGATE of semi-crystallized Silica, with a broad, distinct white band of Chalcedony associated with waxy Agate, probably the *Cerachates* of Pliny. Polished on one side. $3\frac{1}{2}$ by $2\frac{3}{8}$ ins.

Minas Geraes, Brazil.

No. 329. CHALCEDONIC AGATE of a waxy nature. A circle of minute zones of grey and pale blue Chalcedony, nearly surrounded by a very distinct milk-white band of Chalcedony, which forms the numeral **5**. Well polished. $3\frac{1}{4}$ by $2\frac{3}{4}$ ins.

Minas Geraes, Brazil.

A most striking and interesting specimen, the numeral 5 in the distinct white layer of Chalcedony illustrating well the deposition of silica in minute fortification and concentric zones.

* *Lib.* xxxvii., 54.

No. 330. CARNELIAN AGATE, the *Sardachates* of Pliny. Amorphous centre of Silica with zones of Chalcedony, surrounded by a band of amorphous Silica with a reddish border. Brownish-red by transmitted light. Polished on both sides. Section cut from centre of a pebble, and exhibits exterior. 4⅝ by 3¼ ins.

Banda, India.

No. 331. CARNELIAN AGATE, large section, probably the *Sardachates* of Pliny, consisting of a grey and reddish centre, surrounded by a broad stratum of white Chalcedony, which is further encircled by a broad band of especial interest, as in parts near the centre of the Agate, top and bottom, well-defined strata of red and bluish layers, slightly tilted, are seen, which run into a band of very minute layers, pointing to an interruption during solidification. Round the border also are nodules with transparent centres of Chalcedony, whilst larger ones have cavities. The shape of the original Agate is well shown, the exterior extending all round the section, which was flattened and nearly parallel at the base. A very fine colour by transmitted light. Polished on both sides. 12½ by 5¾ ins.

Minas Geraes, Brazil.

No. 332. CARNELIAN AGATE SECTION. In the centre is a small cavity, with blue and grey Chalcedony, tipped with a thin blood-red, surrounded by an irregular distinct milk-white band embedded in large bands of Sard, with strata of black and white Agate associated with Sard. Fine flammeate colour by transmitted light. Well polished. 9 by 6¼ ins.

Uruguay, South America.

A most interrupted and irregular Agate, exhibiting inlaying and pointing to great disturbance.

No. 333. CARNELIAN AGATE SECTION. The centre is of blue and grey Chalcedony, with a thin blood-red line, surrounded by a distinct milk-white band of Chalcedony embedded in Sard, with strata of black and white Agate associated with Sard. Well polished. 9 by 6 ins.

Uruguay, South America.

No. 334. AGATE, consisting of bluish Chalcedony with red Jasper, which on the exterior shows its stalactitic form. This may be the *Hemachates* * of Pliny, from ἄψα = blood, which was probably a true light-coloured Agate blotched with red Jasper, "blushing with spots of blood"; as says "Solinus (King, p. 207), of which there are very beautiful kinds and not simple red Jasper." Well polished. 3 by 2¼ ins.

Oberstein, Germany.

No. 335. BRECCIATED AGATE, inlaid, red and white, irregularly dispersed with amorphous Silica. Thin and very decided coloured section. Polished both sides. 4 by 2½ ins.

Bandah district, India.

Coll. Dr. Birdwood.

No. 336. AGATE SECTION, very irregular, with pattern broken. A small centre with minute concentric zones surrounded by patchy violet and puce bands, showing great disturbance. Polished on one side. 2¾ by 1½ ins.

Aussig, Bohemia.

No. 337. AGATE SECTION in minute irregular strata of a brownish-red and violet, bordered with translucent Chalcedony. Section of a distorted flattened boulder. Polished. 7 by 3¼ ins.

Galgen-Berg, Oberstein.

No. 338. AGATE, oval, brown, with markings of red Jasper (*Hemachates* of Pliny). Polished. 1½ by 1¼ ins.

Minas Geraes, Brazil.

No. 339. AGATE, consisting of irregular concentric layers of black and grey-white strata. Oval. Small. Polished both sides. Opaque. 2 by 1⅝ ins.

Banda, India.

* *Lib.* xxxvii., 54.

No. 340. CHALCEDONIC AGATE, large oval slab, grey and quite translucent, in which is irregularly embedded a distinct brown opaque Jasper, with white lines of Chalcedony and semi-crystalline Silica. 6⅛ by 3⅝ ins. An important and striking Agate.

Banda, India.

No. 341. AGATE, oval, one side of bluish amorphous Quartz with deep black Agate, artificially coloured ; the other side completely black. Polished. 4 by 3⅛ ins. Cut at Oberstein.

Uruguay.

A very characteristic specimen of artificial colouring of Agates.

No. 342. AGATE. Onyx variety, consisting of a dark centre, surrounded by filiform crypto-crystalline Quartz, further encircled by fortification bands of white and black Chalcedony with amorphous Quartz. Colour artificially deepened. Polished all over. Bevelled edge. 4½ by 3 ins.

Minas Geraes, Brazil.

A beautifully distinct marked Agate.

No. 343. CHALCEDONIC ONYX, oval, a cross-section of, ex-exhibiting three deep distinct strata or bands, consisting of minute layers, one white band between two of bluish Chalcedony. Shaped for cutting a cameo. Thick section, and important stone. 2⅞ by 2⅛ by 1¼ ins. thick.

Cut and coloured at Oberstein, but from Brazil.

No. 344. CHALCEDONIC ONYX, a cross-section forming an Onyx of two deep distinct strata or bands, one white, and the other of bluish Chalcedony. Shaped for cutting a cameo. Thick section. Well polished. 2⅞ by 2⅛ by 1¼ ins. thick.

Cut at Oberstein, but from Brazil.

No. 345. CHALCEDONIC WAX AGATE, portion of. Translucent. Slight fortification. Well polished. 3¼ by 1⅜ ins.

Minas Geraes, Brazil.

I

No. 346. AGATE, Chalcedonic. Dark grey, in irregular concentric lines not well-marked ; with fortification of white Chalcedony. A large specimen, showing exterior. One face polished. 6 by 5 ins.

Uruguay, South America.

No. 347. AGATE, variegated section. Of a violet tinge, also red, stained with oxide of iron, with uneven zigzag and parallel lines. Polished. 5 by 3½ ins.

Campsie Hills, Scotland.

No. 348. CHALCEDONIC AGATE, translucent with centre of botryoidal brown Agate surrounded by patches of black and blue Chalcedony; stained reddish through ferrous oxide. Presenting an irregular patched appearance. Beautiful section by transmitted light Polished on both sides. 11½ by 4¼ ins.

Minas Geraes, Brazil.

No. 349. CHALCEDONIC AGATE, translucent centre brown, botryoidal, encircled by concentric zones of black and blue Chalcedony. Presenting a patched, irregular appearance. Beautiful by transmitted light. Polished on both sides. 10½ by 4 ins.

Minas Geraes, Brazil.

No. 350. AGATE SECTION, brownish-red and black, with centre of blue Chalcedony ; surrounded by concentric bands of brown, blue, and white; reddish near exterior. Well polished. 5¼ by 3 ins.

Minas Geraes, Brazil.

No. 351. ONYX AGATE, pale bluish Chalcedony in minute concentric layers forming a circle. 3⅛ ins. diameter. A crack crosses the centre which has admitted red oxide of iron. Embedded in crypto-crystalline Quartz. 5½ by 5 ins.

Minas Geraes Brazil.

No. 352. ONYX AGATE, rectangular and convex, consisting of brown and black Agate, the colours artificially deepened. Interior a cavity of crystalline Quartz surrounded by well-marked fortifications in white. Cut and polished all over. 5¾ by 4 ins.

Cut at Oberstein, but from Brazil.

No. 353. ONYX AGATE, rectangular slab. Black and brown in fortification strata, colour artificially deepened. Polished all over, edges as well. 6¾ by 3¾ ins.

Cut at Oberstein, but from Uruguay, South America.

A good example of staining.

No. 354. ONYX AGATE, rectangular slab. Black and brown Agate, colour artificially deepened. Cut and polished all over at Oberstein. 5 by 2½ ins.

Minas Geraes, Brazil.

No. 355. AGATE, Onyx variety, semi-crystalline centre with broad bands of black and thin pure white lines of Chalcedony in marked fortification. Colour artificially deepened. Polished on one side. 3⅞ by 2¾ ins.

Cut at Oberstein, but from Uruguay, South America.

No. 356. AGATE, brecciated. Of a most irregular dappled appearance. A creamy-white, with pale bluish-red and brown irregular markings intermingled together. Partly opaque, but exhibits by transmitted light a Sard or Carnelian tint in some parts. 6½ by 5½ ins.

Aussig, Bohemia.

A most instructive specimen, indicating great disturbance during formation.

No. 357. AGATE, stained perfectly black. Rectangular. A fine specimen, illustrative of artificial colouring. Polished on one side. 3⅞ by 2½ ins.

Uruguay, South America.

I 2

No. 358. BRECCIATED AGATE, nearly round, mossiform and pisolitic in appearance. Grey, white, and red Chalcedony, exhibiting a most dappled but extremely interesting structure. Opaque. Polished. 5 ins. in diameter.

District of Aussig, Bohemia

No. 359. Ball of BROWN AGATE in various layers. Diameter 1⅝ ins.

Cut, polished, and drilled at Oberstein, but from South America.

No. 360. Ball of AGATE, in Chalcedony, with broad reddish-brown bands. Diameter 1¾ ins. Cut, polished, and drilled at Oberstein.

Uruguay, South America.

MOSS AGATE.

MOSS AGATE, or Tree Agate, consists of transparent or translucent white or bluish-white Chalcedony, filled with green, brown, red, and black dendritic or moss-looking markings, caused by the deposition of metallic Oxides, probably manganese and iron.

It is mentioned by Pliny under the name *Dendrachates*.* It is used largely for ornamental purposes, such as vases, caskets, tazze, etc., and is one of the most beautiful Agates by transmitted light. Magnificent goblets or vases, nearly a foot high, are in the Earl of Derby's possession. The specimen No. 366 in the Derby Collection is one of the finest slabs known, not only for its beauty but for its abnormal size.

Colours due to Visible Impurities.

No. 361. MOSS AGATE or TREE AGATE. *Dendrachates* of Pliny (from δένδρον = a tree; Book xxxvii., Chap. 54). Large rectangular section of Chalcedony, with green and red moss-like or dendritic forms, caused by the deposition of oxides of manganese and iron, which is distributed through the whole specimen. Red largely disseminated over the specimen, with innumerable white scattered pisolitic spots. Translucent to transparent. Polished both sides. Very beautiful by transmitted light, having the appearance of delicate dense seaweed. 13 by 7¾ ins.

Ahmedabad, India.

An extraordinary and most lovely specimen by transmitted light, forming, with the other examples following, the finest series of Moss Agates known in any collection.

* *Lib.* xxxvii., *cap.* 54.

No. 362. MOSS AGATE. *Dendrachates* of Pliny. Large rectangular section of beautiful moss-green dendrites in bluish Chalcedony. In this specimen there is little red, the deposition being probably solely due to oxide of manganese. A small cavity exists of crystalline Silica. Translucent. Polished both sides. Three edges rough, one polished. 13¼ by 7¼ ins.

Ahmedabad, India.

Another magnificent abnormal specimen, the green moss-like dendritic markings being beautifully disseminated, and most distinct.

No. 363. MOSS AGATE. Large rectangular section of beautifully disseminated green dendritic markings, caused by the precipitation of oxide of manganese embedded in bluish-grey translucent Chalcedony, with irregular white patches of white brown Agate. Polished both sides. 13 by 7¾ ins.

Ahmedabad, India.

A fine thick green specimen with a somewhat dappled appearance of great size like the preceding.

No. 364. MOSS AGATE or TREE AGATE. *Dendrachates* of Pliny. Large rectangular slab of different shades of mossy-green dendritic markings, both light and dark, mottled or spotted white, with greyish-blue Chalcedony, caused by the precipitation of oxide of manganese. Well polished. 13 by 7¼ by ¾ ins. thick.

Ahmedabad, India.

An unusually thick specimen, with the green, light, and dark mossiform markings thickly and beautifully disseminated.

No. 365. MOSS AGATE. The *Dendrachates* of Pliny, which includes both Moss Agate and Mocha Stone. Fine mossy dendrites of a beautiful light and dark green embedded in transparent Chalcedony, with a good deal of red Agate at one side, caused by the precipitation of metallic oxides of manganese and iron. Polished well both sides. Irregular shape. Translucent. 10 by 6½ ins.

Ahmedabad, India

No. 366. Moss Agate or Tree Stone. *Dendrachates* of Pliny (from δένδρον = a tree; Book xxxvii., Chap. 54). Unique section of beautiful green and red mossy-looking dendrites disseminated through a base of bluish white Chalcedony, forming delicate arboraceous markings or lines running into irregular green and reddish patches, caused by the precipitation of oxides of manganese and iron. Semi-crystalline Silica exhibited in places, also white fortifications in bluish Chalcedony. Very beautiful by transmitted light. Translucent. Extremely well polished on both sides. $12\frac{7}{8}$ by $8\frac{1}{2}$ ins., but only $\frac{1}{8}$ of an inch in thickness.

District of Ahmedabad, India.

This remarkable section is unique not only for its loveliness by transmitted light, presenting an appearance of most delicate seaweed in many tints, but for its thinness in relation to its large size. It is a good example of the perfection to which slitting and polishing in France has arrived. It was cut on the Marne by hydraulic power, with diamond powder, and afterwards most beautifully polished on both sides with great care. Its remarkable thinness, only one-eighth of an inch thick, renders it perfectly translucent, suggesting its use for the decorations of windows like stained glass. To advance the suggestion a magnificent plaque, consisting primarily of Moss Agates surrounded by other Agates and Jaspers cut to the required thinness and necessary translucence, was made to the order of the distinguished collector Mr. Alfred Morrison, and has proved a great success, showing that the *hard* stones, with their great beauty and exquisite tints revealed by transmitted light, can now be utilized as well as the softer for window decoration.

No. 367. Moss Agate or Tree Stone. A long rectangular specimen of red and green dendritic mossiform markings in Chalcedony, due to the precipitation of red and green oxide of iron and manganese well and thickly disseminated. Translucent. Polished on both sides. 7 by $3\frac{1}{4}$ ins.

Ahmedabad, India.

This specimen is from the collection of the late Professor James Tennant, F.G.S., by whom it was used for over twenty years to illustrate his lectures at King's College, London, where he was Professor of Mineralogy. It was exhibited in the Exhibition of 1851 in the Indian Department, and was then looked upon as being *unique*, both as to size and beauty. After the Exhibition it passed into Professor Tennant's possession, and at his decease was purchased by the author to place specially in Lord Derby's collection.

No. 368. MOSS AGATE. Rectangular section of a light and dark green caused by dendritic markings of a mossiform character, well disseminated in grey Chalcedony, with here and there patches of white Silica and bluish Chalcedony. By transmitted light the markings are extremely delicate and beautifully reticulated, looking like seaweed. Thin section, well polished on both sides. Extremely beautiful and quite transparent. $7\frac{1}{8}$ by $5\frac{1}{4}$ ins.

Ahmedabad, India.

No. 369. MOSS AGATE. Rectangular section of a light and dark green caused by dendrites looking like seaweed, well disseminated in grey transparent Chalcedony, with patches of white Silica and bluish Chalcedony. The markings are extremely delicate and reticulated. Thin section similar to 368. Well polished on both sides. $7\frac{1}{8}$ by $5\frac{1}{4}$ ins.

Ahmedabad, Punjaub, India.

No. 370. CHALCEDONIC MOSS AGATE. Partly opaque. Bluish Chalcedony, over which is disseminated yellow clouds. A cavity of white crystalline Silica is exhibited on one side, with reddish Agate, associated with a cloudy milky ground, through which runs black lines. Polished on both sides. 11 by $5\frac{3}{4}$ ins.

Bombay district, India.

No. 371. MOSS AGATE. Rectangular specimen. Red mossiform, milky-white and green dendrites from precipitations into Chalcedony of metallic oxides of manganese and iron. Partly translucent. Polished and bevelled. $5\frac{1}{8}$ by $3\frac{1}{2}$ ins.

Punjaub, India.

No. 372. MOSS AGATE. Rectangular specimen. Red mossiform, milky-white and green dendrites from precipitations into Chalcedony of metallic oxides of manganese and iron. Partly translucent. Polished and bevelled. $5\frac{1}{8}$ by $3\frac{1}{2}$ ins.

Paujaub, India.

No. 373. MOSS AGATE, translucent, of a dull reddish colour with green, caused by metallic oxides. Rectangular. Polished, with bevelled edge.   3 by 2¼ ins.

Paujaub, India.

No. 374. MOSS AGATE. An important oval slab. Dense green with light green dendritic mossiform markings, with fortifications of white Chalcedony. Polished at both sides. Cut at Oberstein on the Nahe.   4¾ by 3¾ ins.

Punjaub, India.

No. 375. MOSS AGATE. Dense dendritic green of two shades, with white Agate. Opaque, excepting in parts of the green. Polished one side.   2⅜ by 2½ ins.

Ahmedabad, India.

No. 376. MOSS AGATE. Slice of the same piece, dendritic markings in two shades of green. This section is thinner and translucent. Well polished.   2½ by 2 ins.

Punjaub, India.

No. 377. MOSS AGATE. Dendritic green markings with white Agate. Cut from same piece as the preceding. Well polished.   3 by 2 ins.

Punjaub, India.

No. 378. MOSS AGATE. Green in dendrites, in translucent Chalcedony. Small. Polished all over.   1⅛ by ¾ in.

Punjaub, India.

No. 379. Ball of MOSS AGATE with Jasper. Pinkish tint. Opaque. Diameter 1¾ ins. Cut, polished, and drilled at Oberstein.

Bandah district, N.W. Province of India.

MOCHA STONES.

THESE are specimens of translucent or semi-transparent Chalcedony, in which black, brown, and red dendritical figures are displayed.

The markings look like delicate trees similar to the appearance of dried seaweed, and were supposed for many years to be petrified vegetable matter. They are, however, caused by the precipitation of manganese, iron, and perhaps other metallic minerals. Being of considerable value when fine, imitation Mocha Stones have been made by taking pieces of Chalcedony of the required colour and etching them by the aid of wax, honey, and sulphuric acid, or they are sometimes drawn upon by nitrate of silver, or even marking ink. The forgeries are, however, easily detected, as the tree-like figures are naturally on the surface, whilst the markings upon the true Mochas are often under the surface, and are seen to dip obliquely, and in many cases go right through the stone. They are called Mocha Stones from the first specimens having been found at Mocha in Arabia. They are found largely in *Central* India, but at what precise locality is not known, as the collectors, who make journeys about every ten years for the special purpose of collecting them, are most reticent as to the exact part of the country in which they are found. Mr. John Luff, the eminent Indian engineer, who has made and brought to England the finest collection of Mocha Stones known, some of which are in this collection, informs me that the Indians go into the interior of India for at least six months at a time about every ten years, and return more or less laden with irregular rough unpolished specimens. The Mocha Stone collectors hand down the secret of the locality orally to their children, whom they also instruct to carefully cut, so that the dendritic or tree-like markings appear precisely upon the surface of the stone, neither too faint by under nor over-cutting, and then to polish them by continuous manual labour. The shape chosen 'is generally oval, but sometimes round if an especially good specimen can be preserved by this shape.

No. 380. MOCHA STONE. Oval. The *Dendrachates* or Tree Stone of Pliny. Dendritic Agate. From δένδρον = a tree. Milky-white ground with dark patches of black dendritic markings, with a little brown, due to the precipitation of metallic oxide of manganese and iron. Translucent. Polished. 3½ by 3⅛ ins.

Central India.

No. 381. MOCHA STONE. Oval. *Dendrachates* of Pliny. White Chalcedonic ground of a bluish tinge, with beautiful clear distinct dendritic markings of extreme delicacy, looking like seaweed, caused by metallic oxides. Polished all over. Translucent. 3 by 2¼ ins.

Central India.

Coll. Arthur Wells.

A most perfect and well defined specimen, the dendrites being remarkable for their clear and distinct arborescent appearance.

No. 382. MOCHA STONE. A pure milky-white Chalcedonic ground with fine black and brown dendritic markings of oxide of manganese or iron. Oval. Thin. Translucent. Polished all over. 2⅝ by 2 ins.

Central India.

Coll. John Luff.

No. 383. MOCHA STONE. *Dendrachates*, Pliny. Bluish-white base of Chalcedony with delicate black and brown dendrites. Oval. Polished all over. 2¾ by 1¾ ins.

Central India.

Coll. John Luff.

No. 384. MOCHA STONE. Oval. A bluish-grey Chalcedonic base with thick brown and black dendritic markings of
oxide of manganese.   Polished all over. $2\frac{3}{8}$ by $1\frac{3}{4}$ ins.

Central India·

Coll. John Luff.

No. 385. MOCHA STONE. A white Chalcedonic base
with fine dendritic brown and black arborescent markings.
$2\frac{1}{4}$ by $1\frac{1}{2}$ ins.

Central India.

Coll. John Luff.

No. 386. MOCHA STONE. Oval. Delicate, slender black
and brown dendritic markings well dispersed over a ground
of white Chalcedony.   Rather thick stone.   Polished on face.
2 by $1\frac{1}{2}$ ins.

Central India.

Coll. John Luff.

No. 387. MOCHA STONE. Oval. Delicate black dendritic
delineations upon a base of bluish translucent Chalcedony.
The markings are not quite distinct, as the specimen has
not been correctly cut so as to leave the dendrites at the
top, a thin film of Chalcedony covering them.   Translucent.
2 by $1\frac{5}{8}$ ins.

Central India.

Coll. John Luff.

No. 388. MOCHA STONE. Oval. *Dendrachates* of Pliny.
White and pale bluish Chalcedony with cloudy dendritic or
tree-like markings, caused by oxide of manganese.   Polished
all over.   2 by $1\frac{3}{4}$ ins. long.

Central India.

Coll. A. M. Jacob, of Simla.

No. 389. MOCHA STONE. Oval. Blue Chalcedonic base with black and brown metallic dendritic markings. Translucent. 1½ by 1 in.

Central India.

Coll. A. M. Jacob.

No. 390. MOCHA STONE. *Dendrachates* of Pliny. Blue Chalcedonic ground with delicate brown dendrites, caused by oxide of manganese. Translucent. 1½ by 1¼ ins.

Central India.

Coll. A. M. Jacob.

AGATIZED WOOD.

No. 391. AGATIZED or JASPERIZED WOOD. A characteristic thick section, consisting of a centre of Chalcedony of a bluish tinge partly stained red (probably through iron oxide), occupying the walls and cells of the original wood. The texture, cells, and grain of the wood are distinctly visible. An analysis of a specimen of Silicified Wood, from Oberkassel, made by Brandes, yielded—

Silica ...	... 93·00
Ferric Oxide	... 0·37
Alumina ...	... 0·13
Water ...	... 7·13
	100·63

Well polished. 7 by 6½ ins.

From the Petrified Forest called Chalcedony Park, near Carrizo, Apache, co. Arizona, U.S.A.

The original formation and texture of the wood in the Arizona specimens is clearly discernible, the process of the deposition of the Silica from its solution, infiltrating into the cells of the wood slowly, and gradually assuming the walls of the cells as the wood itself gradually and entirely disappears, leaving Silica in the form of the tree, is well represented.

No. 392. AGATIZED WOOD SECTION. Red Chalcedony with a bluish tinge, occupying the original wall and cells of the tree, exhibiting texture and general wood formation. Well polished. 6 by 5¾ ins.

Petrified Forest of Chalcedony Park, Arizona.

No. 393. AGATIZED WOOD. Chalcedony of a distinct flammeate red, caused by iron oxide infiltrated into the cells and walls, assuming their shape. Exhibits the texture and divisions of the wood distinctly. 5½ by 4½ ins.

Petrified Forest of Chalcedony Park, Arizona.

Striking and highly coloured specimen.

No. 394. AGATIZED WOOD SECTION, with Chalcedonic centre infiltrated into the cells and walls. The bark is particularly well silicified. 4¾ by 4¼ ins.

Antigua, West Indies.

A most characteristic specimen, with the silicified bark in quite its natural shape.

No. 395. AGATIZED WOOD. A fine thick black section with texture and bark distinctly changed into Chalcedony by its infiltration from soluble Silica into the cells and walls. The bark silicified is still quite intact and in its natural shape. Well polished on both sides. 5 by 3½ ins. by 1 in. thick.

Antigua, West Indies.

No. 396. AGATIZED or JASPERIZED WOOD. A thick section distinctly exhibiting the original texture of the tree, although Silica in a soluble form has infiltrated into its cells and walls. 3⅛ by 2½ ins.

Antigua.

No. 397. AGATIZED WOOD. Chalcedony grey, exhibiting close structure, stained slightly red by iron oxide. Infiltrated from soluble Silica, which has assumed the texture of the wood. Portion of section. Polished on one side. 5 by 2½ ins.

Antigua, West Indies.

No. 398. AGATIZED WOOD. Chalcedony grey, exhibiting fine grained texture slightly stained red. Similar to No. 397. Portion of section. Polished on one side. 5 by 2½ ins.

Antigua, West Indies.

No. 399. AGATIZED WOOD. Chalcedony grey, with fine grained texture, slightly stained. Infiltrated into the small close cells from Silica in solution. Similar to Nos. 397 and 398. Polished on one side. Portion of section. 5 by 2½ ins.

Antigua.

No. 400. AGATIZED WOOD, most probably a species of palm. Of a snuffy-brown colour, exhibiting structure very distinctly. Formed also by the infiltration of soluble Silica into the cells of the original wood. A fine and most characteristic specimen, exhibiting the bark in its natural form very well. Polished on one side. 6 by 4 ins.

Antigua.

No. 401. AGATIZED WOOD SECTION, oval, most probably a species of palm. Exhibiting distinctly the texture by Silica from a soluble solution, infiltrated into the cells. Grey ground with reddish markings. Cut and polished on both sides with bevelled edges.

Vindyah Mountains, India.

No. 402. AGATIZED WOOD SECTION. Oval. Chalcedony, assuming by infiltration of soluble Silica, the cells and walls of the original tree, a palm. Grey with reddish spots, due to iron oxide. Cut and polished on both sides with bevelled edges. $3\frac{5}{8}$ by $3\frac{1}{4}$ ins.

Vindyah Mountains, India.

No. 403. AGATIZED WOOD. Exhibiting distinctly texture and formation of cells and walls from the infiltration of soluble Silica ; running from a blackish to a grey colour. $3\frac{1}{2}$ by 3 ins.

Antigua, West Indies.

ONYX.

ONYX ('Ονύχιον of Theophrastus) is derived from the Greek word ὄνυξ=a finger-nail; mentioned several times by Pliny, Lib. xxxvii., cap. 24, who applied it to many varieties of Chalcedony. Onyx is really a variety of Agate occurring in straight, even parallel planes or concentric zones of well-defined colours, one of which must be *white* associated with black, brown, grey, or any other alternate colour. "Onicolo," or "Nicolo," is a term given to a black onyx with a thin layer of white over it, which has a bluish tinge, caused by the black showing through, and is used for cutting Intaglie and Camei. When the band is white, with one or more red bands, it is called *Sard*-onyx.

A noted ancient work of art cut from an onyx is the Mantuan Vase at Brunswick. It represents a cream-jug about 7 inches high and 2½ broad, and is cut from a single stone.

———

No. 7. Onyx.

No. 404. ONYX. Consisting of a pure black centre, with distinct white and brown concentric zones of white Chalcedony. A most characteristic example, with artificially deepened colours. Well polished. 4 by 3¾ ins.

Uruguay, South America.

No. 405. ONYX. Black circle of Agate, surrounded by white Chalcedonic concentric zones. Colour artificial. 3⅞ by 3 ins.

Minas Geraes, Brazil.

No. 406. ONYX. Consisting of a centre of brownish-black Agate, with milky-white concentric zones embedded in crypto-crystalline Silica. Polished on one side. 3½ by 2¾ ins.

Minas Geraes, Brazil.

K

No. 407. ONYX. Small specimen, consisting of black Agate with white layers of Chalcedony in zigzag formation. Polished on one side. 2 by $1\frac{7}{8}$ ins.

Brazil.

No. 408. ONYX. Consisting of two distinct circles, looking like the eyes of an owl: one circle formed of a centre of *brown* Agate with white concentric zone of Chalcedony ; the other a centre of *white* Chalcedony, with brown zone of Agate, surrounded by white distinct layers of Chalcedony. Both circles polished. $2\frac{1}{4}$ by 2 ins.

Cut at Oberstein, but from Brazil.

A striking example. The two circles are cut at different angles.

No. 409. ONYX. A cross stalactitic section. A perfect and distinct circle of minute concentric zones, the centre formed of brownish Agate, surrounded by black and brown zones, most perfectly and regularly arranged. Diameter 3 ins. Artificially coloured at Oberstein on the Nahe.

Minas Geraes, Brazil.

A most beautiful and characteristic specimen of regular deposition of concentric Silica. These concentric zones are not produced by cutting the Agate in the usual flat way, but are really cross sections of layers of *stalactitic* Chalcedony.

No. 410. ONYX. A cross stalactitic section in black and brown concentric circles, the outer of white Silica. Reddish by transmitted light. Polished well on both sides. Colour artificially deepened at Oberstein, but from Brazil. Diameter $2\frac{1}{4}$ ins.

Minas Geraes, Brazil.

No. 411. ONYX. A cross stalactitic section in most beautiful concentric layers or zones. An irregular centre, blackish, with white thin layer of Chalcedony, and deep brownish band, surrounded by thin white zones with a brown Carnelian band outside. Polished both sides. Diameter $2\frac{1}{2}$ ins.

Bandah, India.

A remarkably beautiful cross section cut from a stalactite of Chalcedonic onyx.

No. 412. ONYX. Cross section cut from a stalactite of Chalcedony in most regular and beautiful concentric zones. A whitish centre surrounded by a layer of brown and thin successive zones of white and brown Chalcedony ; and further encircled by a deep brown Sardonyx layer, with white zones on the exterior. Diameter 2⅜ ins.

River Ken, Bandah, N.W. Province of India.

A lovely and most perfect specimen, illustrating the wonderful symmetry and perfect deposition of concentric layers of Silica. One of the first of these stalactite sections was introduced into England at the Exhibition in 1862. It was about three inches in diameter, of white and dark brown Sardonyx, and was utilized as a lid for the top of a solid gold tripod, most elaborately chased and jewelled.

No. 413. ONYX. Another cross section of a stalactite, consisting of a grey centre, with beautiful white and brown alternate concentric zones, with a Carnelian exterior band. Polished on both sides. 2¼ ins. diameter.

River Ken, Bandah, India.

Another lovely and most characteristic example of regular contentric alternate depositions of Silica. A perfect gem.

No. 414. ONYX, of a milky and bluish white. Cross section cut from a Chalcedonic stalactite. A grey centre surrounded by a bluish chalcedonic zone, with further circles of milky-white Chalcedony and Agate. Well polished on both sides. 2½ ins. diameter. India.

No. 415. ONYX. Principally white. Cross section cut from a stalactite, consisting of a white centre with thin brown and white alternate zones. Well polished on both sides. 2½ by 2¾ ins. India.

No. 416. ONYX (two specimens), consisting of black centres with distinct milky white layers. Variety termed " Onicolo " or " Nicolo," used for engraving Camei and Intaglie. Colour artificially deepened. Polished all over. 1½ by 1¼ ins.

Cut at Oberstein, but from Brazil.

The Agate termed " Onicolo " or " Nicolo " consists generally of two strata ; the upper has the subject engraved *through* it, generally being ground very thin—not much thicker than a card, so that the design is seen of a bluish-white *outside*, and black through the white layer *within*.

K 2

No. 417. ONYX. A brownish centre, with a distinct milky-white layer. Variety termed "Onicolo," similar to No. 416. Used for the engraving of Camei and Intaglie. 1⅞ by 1½ ins.

Cut at Oberstein, but originally from Brazil.

No. 418. Ball of ONYX. Agate in black and white, concentric layers. Diameter 1¾ ins. Stained, cut, polished, and drilled at Oberstein.

Uruguay, South America.

No. 419. Ball of ONYX. Black with white irregular layers. Diameter 1½ ins. Stained, cut, polished, and drilled at Oberstein.

South America.

No. 420. Ball of Chalcedonic ONYX. Grey and pale blue colour. Diameter 1¾ ins. Stained, cut, polished, and drilled at Oberstein.

Minas Geraes, Brazil.

No. 421. Ball of ONYX. Exceptionally beautiful, consisting of translucent Chalcedony, stained yellow, with a broad band of white. Diameter 1¾ ins. Stained, cut, polished, and drilled at Oberstein.

Minas Geraes, Brazil.

SARDONYX.

SARDONYX, from the Greek of Σάρδιον = sard, and ὄνυξ = a nail; mentioned several times by Pliny (see Lib. xxxvii., cap. 23). It is a variety of Onyx in layers, one of which must be white and the other reddish-brown, which is termed Carnelian or Sard.

A great deal of confusion exists about the terms Sardonyx and Onyx, and it may be as well here to correct a popular error that Onyx means a stone of two strata, and Sardonyx of three or more. The terms have no reference in any way to the number of strata. Onyx is a stone which presents the superposition of at least one stratum over another, one being of necessity *white*, whilst the other may be black, brown, grey, or any other colour; but if that colour be *sard*, it constitutes a Sardonyx (Sardonyx, *candor* in *Sarda*; Pliny, Lib. xxxvii.); and there may be three or even four or five strata of either Onyx or Sardonyx. The French word, *Sardoine*, so nearly approaching Sardonyx, causes at times confusion.

No. 8. Sardonyx.

No. 422. SARDONYX. Section of a flammeate red Sard; irregular circle surrounded by a most distinct white band of Chalcedony, which extends right and left at the base, with crystallized Silica. 4¼ by 3¾ ins.

Sandeopoldo, Rio Grande do Sul, Brazil.

A striking specimen, the vivid colour of which has been obtained undoubtedly by artificial means.

No. 423. SARDONYX. Rectangular section, whitish filiform Silica in the centre, with deep bands of red and thin white embedded in bluish Chalcedony. Polished on one side. 3⅞ by 2½ ins.

Cut at Oberstein, but from Rio Grande do Sul, Brazil.

No. 424. SARDONYX. Thin rectangular slab of faint red Sard, and white Chalcedony, alternate bands. Exhibits the characteristics of Sardonyx well. Translucent. 4½ by 3½ ins.

Cut at Oberstein, Rio Grande do Sul, Brazil.

No. 425. SARDONYX. Rectangular slab of faint red Sard and white Chalcedony, alternate bands, similar to No. 424, but a thinner section. Translucent. 4½ by 3½ ins.

Rio Grande do Sul, Brazil.

No. 426. SARDONYX. Small section, in faint red Sard and white alternate bands. Translucent. Well polished. 3½ by 2⅞ ins.

Rio Grande do Sul, Brazil.

No. 427. SARDONYX. Rectangular slab, in faint red and white alternate bands. Translucent. 4½ by 3½ ins.

Cut at Oberstein, but from Brazil.

No 428. SARDONYX. Oval. Of several strata or layers. White Chalcedony and brown Sard. Cut for engraving a cameo. Polished all over. 1⅞ by 1¼ ins.

India.

No. 429. SARDONYX. Flammeate red and brown Sard, with an aventurine structure. Thick rectangular specimen, with corners removed. Cut for a paper-weight. The sides exhibit distinct parallel strata of red, white, brown, and grey, dappled in appearance ; colour deepened artificially at Oberstein. Bevelled edges. Polished all over. 6 by 3½ ins.

Minas Geraes, Brazil.

No. 430. SARDONYX. Cut in the shape of a long round column, thicker at the bottom than the top ; flammeate red Carnelian alternate with grey Chalcedonic layers. Polished all over. Made at Oberstein ; colour artificially deepened. Well polished. 10½ ins. long.

Minas Geraes, Brazi'

No. 431. SARDONYX. Oval. Flammeate red Sard in white Chalcedony. Cut and polished all over. 2 by 1¾ ins.

Campsie Hills, Scotland.

No. 432. Ball of SARDONYX. Top in white, with alternate bands of red Sardonyx and white Chalcedony. Diameter 1¾ ins. Stained, cut, polished, and drilled at Oberstein.

Minas Geraes, Brazil.

No. 433. Ball of SARDONYX of red and white alternate layers of Sard and white Chalcedony. Diameter 1¼ ins. Stained, cut, polished, and drilled at Oberstein.

Minas Geraes, Brazil.

No. 434. Ball of SARDONYX. Consisting of grey alternate layers of Chalcedony and Sard, with white Chalcedony in layers at the top. Diameter 1¼ ins. Stained, cut, polished and drilled at Oberstein.

Brazil.

No. 435. Ball of SARDONYX. In zigzag faint red and white bands. Diameter 1¾ ins. Stained, cut, polished, and drilled at Oberstein.

Minas Geraes, Brazil.

No. 436. Ball of RED SARDONYX. Faint white bands with red. Diameter 1½ ins. Stained, cut, polished, and drilled at Oberstein.

Brazil.

AGATE JASPER.

No. 9.   Agate Jasper.

No. 437. AGATE JASPER. Of a bluish and brown colour, associated with Chalcedony, and exhibiting indications of a silicified coral or sponge.   4 by 2½ ins.

Antigua, West Indies.

SILICEOUS SINTER.

No. 10.   Siliceous Sinter.

No. 438. SILICEOUS SINTER. A grey, irregular honey-comb Quartz, a Silica formed by the deposition of water, containing an appreciative amount of soluble Silica in solution. Sometimes called Geyserite.   An analysis of a specimen of Geyserite from the same locality (Iceland), made by Forchhammer yielded :

Silica	84·43
Ferric Oxide	1·91
Alumina	3·07
Lime...	0·70
Water	7·88
Soda and traces of Potash	0·92
Magnesia	1·06
	99·97

1¾ by 1¼ ins.

Vatna Jökull, Iceland.

FLINT.

No. 11. *Flint.*

No. 439. FLINT. *Lapis vivers* and *Silex pt.,* Pliny. Feuerstein of the Germans. La pierre a feu, *Fr.* Somewhat allied to Chalcedony, but opaque and dull. Black to grey, exhibiting a distinct and deeply conchoidal fracture with a sharp cutting edge. According to Fuchs, the silica of flint is partly soluble. Generally contains water, with a small amount of ferric oxide and alumina. An analysis made by Von der Mark of a specimen of flint yielded :

Silica	95·18
Oxide of Manganese ...	0·15
Lime...	0·78
Trace of Potassium and Soda	0·08
Water	4·00
Ferric Oxide, a trace of	
	100·19

4½ by 2 ins.

Isle of Portland.

One of the commonest, but most useful, forms of Silica allied to Chalcedony, owing its colour generally to carbonaceous matter. The material used by Palæolithic or præhistoric man for the manufacture of his first implements, axes, chisels, gouges, and hammers,* and used, before the introduction of percussion locks and lucifer matches, for making into gun flints, and with a piece of steel for igniting timber, etc. It is extensively used also in Kent and Sussex for making roads, as well as in Cumberland and many other northern counties for the building of houses and walls. Found largely in irregular shaped nodules arranged in layers in the chalk formation.

* *See a Lecture upon "Primeval Man," illustrative of the præhistoric remains in the Ethnographical collection of the Liverpool Museum, by the late Sir J. A. Picton, F.S.A., Chairman of the sub-committee of the Liverpool Museum.*

HORNSTONE.

No. 12. *Hornstone.*

No. 440. HORNSTONE. A variety of flint, but of a more brittle and splintery fracture. A fine white specimen, with dendritic or tree-like markings, caused by oxides of manganese and iron, tinged slightly red at one side. Well polished on both sides. 5 by 3½ ins.

Altai Mountains, Asiatic Russia.

Hornstone is the name given to an impure flint with a more or less flat fracture, not breaking with a conchoidal one like flint.

No. 441. HORNSTONE. Brown with grey-white spots. Polished on one side. 6¼ by 4¾ ins.

Bandah, India.

TOUCHSTONE or LYDIAN STONE.

No. 13. *Basanite. Lydian or Touchstone.*

No. 442. BASANITE. TOUCHSTONE. LYDIAN STONE or LYDITE. Of a velvet-black. Not splintery. Passes, according to Dana, into varieties of a flinty rock of different colours, particularly grey. Called a siliceous Slate, also *Phthanyte.* Very hard. Unpolished. Opaque. 3 by 2¼ ins.

Langenstriegis, Freiberg.

A well-known stone, used by jewellers principally for testing the purity of gold. The metal is rubbed on the stone, leaving a certain tint or colour, which when touched with acid varies according to its purity, indicating distinctly the quality of the alloy. It is the *Basanites* of Pliny (Book xxxvi., Chap. 11), who also mentions it under the name of *Coticula,* literally whetstone, as well as *Heraclian* and *Lydian* stone (Book xxxiii., Chap. 43). According to Theophrastus, "this stone was nowhere to be found, except in the river Tinolus (in Lydia) but at the present it is found in numerous places." Pliny further says the Egyptians discovered it in Ethiopia, and that it resembled iron in colour and hardness, whence its name, Βάσανος = a touchstone : "*Invenit eadem Ægyptus in Æthiopia, quem vocant basalten, ferrei coloris atque duritiæ. Unde et nomen ei dedit.*" He mentions a large block of it used for a group, dedicated by the Emperor Vespasian in the Temple of Peace, which represented the river Nilus with sixteen children sporting around it. He mentions another large block in the temple of Serapis, at Thebes. forming the Statue of Memnon (the Amenophis of the Egyptians), a monarch of the second dynasty. A large statue, still to be seen at Medinet Abou, on the Libyan side of the Nile, in a sitting posture, about sixty feet high, is probably the statue to which Pliny referred.

JASPER.

JASPER, or Iaspis of Pliny, a name by which the true Jasper and many other varieties were known to the Greeks, is a Silica occurring in masses, seldom in strata. It contains some clay and yellow or red iron oxide ; the red is the anhydrous oxide, whilst the yellow is the hydrous. If the yellow be subjected to heat the water is driven off and it becomes red. It was well known to the ancients, and is often mentioned in the Bible, being the twelfth stone of the breast-plate of the High Priest (Ephraim). Pliny describes several varieties, "red as Hæmatitis" (Lib. xxxvii., 60), the best coming from Scythia, Cypria, and Egypt, on the banks of the Nile. It is said the Column of Memnon and the foundation of Pompey's statue were made of Jasper, and in the ruins of Herculaneum and Pompeii many fragments have been found.

It is opaque. The colours are brown, black, red, white, green, and blue, and it is generally classified into red, brown, dark green, greyish-blue, blackish or brownish-black striped or *riband* Jasper, Egyptian Jasper, Jasponyx, and Jasperized Wood.

The EGYPTIAN JASPER is a peculiar stone which occurs in spheroidal pebbles, of a grey-brown and red colour, which, when cut and polished, exhibits annular markings round the centre, conoidic, or takes grotesque shapes, sometimes assuming the shape of faces and heads, causing it to be called "Face" stone. An analysis of Jasper yielded Beudant :

Silica ...	...	93·57
Ferric Oxide	...	03·98
Alumina ...	...	0·31
Lime... ...	...	1·05
Water ...	...	1·09
		100·00

No. 14. *Jasper.* *Brown or Ochre - Yellow.*

No. 443. JASPER. Iaspis of Pliny. Very large specimen, principally brown or ochre-yellow, in map-like markings, tinged violet in some places and at others black towards
the border. Irregular in pattern, unusually large and thick
specimen. Well polished. Opaque. 13 by 8 ins. and ¾ in.
wide.

Siberia.

An impure Quartz, coloured ; the brown of this specimen being
derived from hydrous ferric oxide, which, if heated, becomes
red through the evaporation of the water.

This coloured Jasper is most probably a variety mentioned by Pliny
as existing under the name of *Terebenthine coloured Iaspis—
Yellow Jasper* according to Ajasson.* Pliny says : " *Item tere-
binthizusa, improprio (ut arbitror) cognomine, velut e multis
eiusdem generis composita gemmis*," that is to say, it is improperly so-called, "being composed of numerous gems of this
description," viz., yellow and other coloured Jaspers, which would
apply to the above better than a purely plain yellow specimen.

No. 444. JASPER. In part Agate, of mossy-looking description, yellow and reddish ground, associated with pale blue
Chalcedony. Opaque, and in parts translucent. Polished on
both sides. An irregular-looking but vivid coloured and interesting specimen. 9½ by 6 ins.

Ekaterinburg, Siberia.

Red Hæmatitis.

No. 445. JASPER. Unique oval specimen, of a bright
blood-red colour, associated with green dendritic Moss
Agate (examine by transmitted light), which is translucent.
Highly polished on both sides. 10½ by 6 ins.

District of Bandah, India.

Coll. Arthur Wells.

A magnificent and vivid blood-red coloured specimen of unusual size.
The association of an opaque Jasper of this description with the
translucent Moss Agate is extremely rare.

No. 446. JASPER. Rectangular, of a continuous bright
terra-cotta colour. Most beautifully polished on one side.
Opaque. 8 by 5½ ins.

India.

A most characteristic example of a single vivid-coloured Jasper.

* *King interprets it as "Yellow, like turpentine."*

Red Hæmatitis.

No. 447. JASPER. A dark blood-red variety, streaked with black lines and mottled, embedding large sage green patches of probably Olivine. Has a dappled appearance in some parts. Well polished on one side. 8 by 6½ ins.

Siberia.

No. 448. JASPER. Of a deep red-mottled colour, with yellow, pale blue Chalcedony intervening, partly pisolitic. Opaque. Polished on both sides. 6¼ by 6 ins.

Ekaterinburg, Siberia.

No. 449. JASPER. Variegated mottled red and brown, in irregular map-like markings, giving it a dappled appearance, with faint fortification; blue Chalcedony intervening. Opaque. Polished on both sides. 8 by 5 ins.

Ekaterinburg, Uralian Mountains, Siberia.

No. 450. JASPER. Variegated. Cut as a paper-weight. Red white, and brown mottled appearance. A thick specimen, with bevelled curved edges. Polished on one side. Opaque. 6½ by 4½ ins.

Ekaterinburg, Siberia.

No. 451. JASPER. Brown and red, in irregular semi-strata, or layers. Somewhat contorted. Polished on one side. Opaque. 6¼ by 5 ins.

Baden-Baden.

No. 452. JASPER. Principally red with stripes of yellow. A rather thick specimen. Opaque. Well polished. 6 by 3 ins.

Ekaterinburg, Siberia.

No. 453. JASPER. A long rectangular specimen. Yellow, associated with red, of a dappled appearance. Opaque. Polished all over. With bevelled edges. 7½ by 2½ ins.

Ekaterinburg, Siberia.

Coll. Arthur Wells.

No. 454. JASPER. Of a dappled appearance. Yellow, with red. A long rectangular specimen. Polished all over, with bevelled edges. Opaque. $7\frac{1}{8}$ by $2\frac{1}{2}$ ins.

Ekaterinburg, Siberia.
Coll. Dr. Birdwood.

No. 455. JASPER, variegated. Rectangular, with a mossiform appearance. Yellow, red, and also of a violet tinge ; somewhat dappled. Polished on one side. Opaque. 4 by 3 ins.

Minas Geraes, Brazil.
Coll. Arthur Wells.

No. 456. JASPER. Rectangular, of a vivid blood-red colour, having a somewhat banded appearance. Opaque. Beautifully polished and mounted on an iron base. $4\frac{3}{8}$ by $3\frac{1}{4}$ ins.

Ekaterinburg, Siberia.

No. 457. JASPER. Cut from an irregular boulder of a dark blood-red colour. Polished. Opaque. $4\frac{1}{2}$ by $2\frac{3}{8}$ ins.

Banda, India.

No. 458. JASPER. Section cut from a distorted amygdaloidal shaped pebble. Sandy yellow in the centre, with a terracotta red border. Polished. 5 by $2\frac{1}{2}$ ins.

Banks of the Nile, Egypt.

No. 459. JASPER. Irregular shape. Red, mottled with brown, with Chalcedony of a bluish tinge intervening. Opaque. $4\frac{3}{4}$ by $2\frac{1}{2}$ ins.

Banda, India.

No. 460. JASPER. Irregular specimen with a yellow ground, with violet and white coloured Chalcedony. Dappled appearance. Opaque. $5\frac{1}{4}$ by $3\frac{1}{4}$ ins.

Phrygia.

No. 461. JASPERY-AGATE. Large oval specimen, associated with Hornstone. Milky-white, stained with a large patch of red oxide of iron, exhibiting irregular Chalcedonic layers. Spotted. Thick specimen. Polished. 6⅜ by 4 ins.

Banda District, India.

Coll. Arthur Wells.

No. 462. JASPER. A very large flammeate red coloured specimen, with white pisolitic spots irregularly disseminated. It has also a crystalline cavity which pierces the section. Well polished. Opaque. 14 by 8 ins.

Ekaterinburg, Siberia.

An exceptionally large, important, and distinctly coloured example, of a mossiform and somewhat dappled appearance.

No. 463. JASPER. Light brown, with an irregular centre ; of a pinkish hue. A thick section, cut from a boulder. Opaque. Well polished. 5¾ by 4 ins.

Near Cairo, Egypt.

No. 464. JASPER. Light brown colour, with irregular centre of a pinkish colour. Similar to No. 463, but a thinner section. Opaque. Well polished. 5¾ by 4 ins.

Near Cairo, Egypt.

No. 465. JASPER. Light brown hue, with a pinkish tinge in the centre. Similar to Nos. 463 and 464. Opaque. Well polished. 5¾ by 4½ ins.

Near Cairo, Egypt.

No. 466. JASPER. Rectangular, of a brick-dust colour, finely pisolitic in character. Opaque. Polished. 4⅜ by 3¼ ins.

Ekaterinburg, Siberia.

No. 467. JASPER. Irregular shape, with a flammeate red, light, and dark colour. Opaque. Polished. 3¼ by 2½ ins.

Banda, India.

No. 468. JASPER, of a blood-red colour. Circular and opaque. Cut from a water-worn pebble. 2¾ by 2½ ins.

Plains of Argos.

The Plains of Argos are strewn with thousands of round and nodular pebbles of this description and colour, which may be taken as a characteristic example.

No. 469. JASPER. Red in matrix of quartzite. A small square, opaque specimen. 2 by 1¾ ins.

Baden-Baden.

No. 470. JASPER. A vivid deep-red, with streaks of blue and yellow Chalcedony and Agate in fortifications. Cut somewhat of a heart shape, with deep bevelled edge. Well polished. Opaque. 4½ by 3½ ins.

Ekaterinburg, Siberia.

A striking and beautiful specimen of Jasper, of most decided colour.

No. 471. JASPER, rectangular. Red. Partly banded. Polished on one side. Opaque. 2½ by 1½ ins.

Ekaterinburg, Ural Mountains, Siberia.

No. 472. JASPER, rectangular. Of an unusual slate colour, with veins. Opaque. 3⅝ by 2⅜ ins.

Cut and polished at Oberstein, but from Siberia.

No. 473. JASPER, rectangular. Yellow, with irregular red streaks; partly banded, with whitish clouds. 3 by 1¾ ins.

Banda, India.

No. 474. JASPER, rectangular. Yellow and mossiform in appearance. Bevelled edges. Polished all over. Opaque. Small. 2¾ by 2 ins.

Bandah, N.W. Province of India.

L

No. 475. JASPER, brown mottled with brecciated white agate. Pisolitic in appearance. Rectangular, with bevelled edge. Polished all over. Opaque. $3\frac{1}{4}$ by $1\frac{3}{8}$ ins.

Bandah, N.W. Province of India.

No. 476. JASPER, oval. Deep black colour, with whitish scattered cloud-like interruptions. Polished on both sides. $3\frac{1}{4}$ by 2 ins.

Banda, India.

No. 477. JASPER, white, with red of a mottled appearance in semi-fortification bands. Polished all over. Opaque $3\frac{1}{2}$ by $2\frac{1}{2}$ ins.

Baden-Baden.

No. 478. JASPER, rectangular. Red and brown, with dark brecciated markings. An irregular specimen. Polished all over with bevelled edges. Opaque. $3\frac{1}{4}$ by $1\frac{7}{8}$ ins.

Bandah, N.W. Province of India.

No. 479. JASPER, of a creamy-brown conoidic form, consisting of concentric layers of a creamy-white and light brown. Well polished, and exhibits concentric zones of Jasper very well. $4\frac{1}{4}$ by $3\frac{3}{8}$ ins.

Near Cairo, Egypt.

No. 480. JASPER, of a flammeate red, with darkish bands and well polished. Opaque. 4 by $2\frac{3}{4}$ ins.

Ekaterinburg, Siberia.

No. 481. MOTTLED JASPER, rectangular, of a reddish and brownish mossiform appearance, with white. Polished. Opaque.

Cut at Oberstein, but from Uruguay.

No. 482. JASPER, a deep red, with white irregular lines and patches, with blue Chalcedony intervening. Polished on one side. 4 by 3 ins.

Bandah, India.

No. 483. JASPER, a deep red, with white irregular lines and patches, with Chalcedony intervening. Polished on one side. 4 by 3 ins.

Ekaterinburg, Siberia.

Banded or Riband Jasper.

No. 484. JASPER, banded, Riband Jasper. Band-Jaspis of the Germans, in broad horizontal bands of reddish-brown with sage-green. 4½ by 3¼ ins.

Ekaterinburg, Ural Mountains, Siberia.

A most distinctive and characteristic specimen of Riband Jasper.

No. 485. JASPER, banded. Riband Jasper in black and white Hornstone-looking parallel bands. Cut from the interior of a pebble and exhibiting its form by the outside edge. 3¼ by 2 ins.

Ekaterinburg, Siberia.

No. 486. JASPER, banded, in broad distinct parallel bands of a creamy-white and terra-cotta red. 3½ by 3 ins.

Kundravinsky, Sloboda, Troitsk, Russia.

No. 487. JASPER, banded (Riband Jasper). In parallel bands of a liver-red and whitish-grey. A smooth and porcelain looking Jasper. Opaque. 2¾ by 2¼ ins.

Banks of the Nile, Egypt.

No. 488. JASPER. An irregularly marked specimen with a deep brown base and creamy-white irregular bands. Polished. 3½ by 2½ ins.

Ekaterinburg, Ural Mountains.

L 2

No. 489. JASPER, half of a pebble. In irregular bands of brown and red. One side polished. $3\frac{3}{8}$ by $2\frac{1}{8}$ ins.

Plains of Argos.

No. 490. JASPER, rectangular, in broad brown and blackish bands. One side polished. 4 by 2 ins.

Schlottwitz, Saxony.

No. 491. JASPER. Brown, of a mottled appearance with black spots disseminated. Polished. $4\frac{1}{8}$ by $3\frac{3}{4}$ ins.

Near Cairo, Egypt.

No. 492. JASPER, oval. Brown with red patches. Small. $1\frac{1}{2}$ by $1\frac{1}{4}$ ins.

Freiberg District, Saxony.

No. 493. JASPER, water-worn pebble. Red. Partly polished, following the natural shape of the stone. $2\frac{1}{2}$ by $2\frac{1}{2}$ ins.

Plains of Argos.

A most characteristic specimen of the pebbles which are strewn by thousands over the Plains of Argos.

No. 494. JASPER, red and green, associated with translucent Heliotrope.

The slab is spotted with spheroids and circles of Chalcedony, which are probably produced by alteration. These spheroids, which are of white, blue, and reddish tints, are very interesting, as they are white, red, or blue on one side of the section and *vice-versâ* on the other, although the specimen is only the eighth of an inch in thickness. In some parts it is of a mossiform character, approaching the Moss Agate, whilst in others it is a translucent Heliotrope. Opaque in some parts and translucent in others. Cut from an irregular specimen, and well polished on both sides. 6 by $5\frac{1}{2}$ ins.

Berezov, Siberia.

A remarkably formed specimen of exceptional interest, inviting a great question as to its extraordinary and varied formation as well as to the change of colours in the spheroids and circles.

No. 495. JASPER, red and green, associated with translucent Heliotrope.

The slab is spotted with spheroids and circles of Chalcedony, which are probably produced by alteration. These spheroids, which are of white, blue, and reddish tints, are very interesting, as they are white, red, or blue on one side of the section and *vice-versâ* on the other, although the specimen is only the eighth of an inch in thickness. In some parts it is of a mossiform character, approaching the Moss Agate, whilst in others it is a translucent Heliotrope. Opaque in some parts and translucent in others. Cut from an irregular specimen, and well polished on both sides. $6\frac{1}{2}$ by 5 ins.

Berezov, Siberia.

A remarkably formed specimen of exceptional interest, inviting a great question as to its formation and the change of colours in the spheroids.

No. 496. JASPER turning into wood. Grey body with a darkish border. A small section. $1\frac{1}{2}$ ins. in diameter.

Antigua, West Indies.

No. 497. JASPER, of a deep brown, with a lightish brown band in the centre. Exhibits exterior of stone. $2\frac{1}{4}$ by $1\frac{7}{8}$ ins.

Desert of Cairo, Egypt.

No. 498. JASPER, of a brecciated description, associated with Hornstone. Grey-white ground with yellow and black patches indiscriminately disseminated over it. Large oval specimen. $6\frac{3}{4}$ by $4\frac{7}{8}$ ins.

Bandah, India.

Coll. Arthur Wells.

No. 499. JASPER, banded. Riband Jasper. In broad layers of creamy-white and liver-red Silica. Well polished. $1\frac{3}{8}$ in. square.

Banks of the Nile, Egypt.

Egyptian Jasper.

No. 500. EGYPTIAN JASPER. Half of a rough pebble, showing peculiar conoidic form in brown and yellowish irregular concentric layers. Called also "Face stone" from black lines which often assume the form of the human face. Exterior of stone exhibited. 4½ by 2¾ ins.

Banks of the Nile, Egypt.

A celebrated "Face specimen" in Egyptian Jasper is in the British Museum—a portrait, more or less distinct, of Chaucer being represented on it. Portraits, the form of animals and fish, are not infrequent in Agates and Jaspers, and are termed generally "lusus naturæ." Two Agates, one representing very distinctly the head and neck of a giraffe and the other a sea lion, were lately in the possession of the author, but now are in the collection of J. E. Hodgkin, Esq., of Richmond, who possesses a unique collection of such specimens.

No. 501. EGYPTIAN JASPER. Exhibiting the peculiar conoidic or spherical forms in brown and yellowish tints, which are characteristic of Egyptian Jaspar. Rectangular, with bevelled edges. Opaque. Polished all over. 4 by 2¼ ins.

Banks of the Nile, Egypt.

No. 502. EGYPTIAN JASPER. Pebble, with one portion polished. Brown and yellowish. Exterior of pebble exhibited. Opaque. 3 by 2⅛ ins.

Banks of the Nile, Egypt.

No. 503. JASPER, brecciated. Brown and yellowish. Inlaid, showing conoidic form. Edge, dark brown. Opaque. Well polished. 3 by 2½ ins.

Banks of the Nile, Egypt.

No. 504. EGYPTIAN JASPER. Variety termed "Face stone," in brown and yellowish spheroidal or conoidic forms. Polished on both sides. Opaque. 2¾ by 2½ ins.

Banks of the Nile, Egypt.

No. 505. EGYPTIAN JASPER. Section in brown and yellowish, exhibiting conoidic form, with deep brown border. Opaque. 2¾ by 1¾ ins.

Banks of the Nile, Egypt.

No. 506. JASPER, brecciated, of a mossiform description, looking somewhat like Moss Agate. Brown and white. Rectangular with bevelled edges. Polished all over. 2 by 1½ ins.

Bandah, India.

No. 507. JASPER. Pure and cloudy specimen. White with brown patches disseminated here and there. Opaque. Polished on one side. 4¾ by 3½ ins.

Bandah, N.W. Province of India.

No. 508. JASPER, half of a green boulder largely impregnated with iron, with brownish bands and spots of the same. Smooth exterior. 5½ by 3¾ ins.

Ekaterinburg, Siberia.

No. 509. JASPER, of a pinkish tint. Exhibits exterior of stone. Polished on one side. 3¼ by 2¼ ins.

Baden-Baden.

No. 510. Ball of WHITE JASPER, with yellowish and reddish markings. Diameter 1½ ins. Cut, polished, and drilled at Oberstein.

Ekaterinburg, Siberia.

No. 511. Ball of JASPER of a light homogeneous red colour. Diameter 1¾ ins. Drilled, cut, and polished at Oberstein.

Ekaterinburg, Siberia.

No. 512. Ball of JASPER. Blood-red, partly striped. Diameter 1¾ ins. Drilled, cut, and polished at Oberstein.

Ekaterinburg, Siberia.

No. 513. Ball of JASPER. Black, with large white patches and whitish spots. Diameter $1\frac{1}{2}$ in. Cut, polished, and drilled at Oberstein.

Ekaterinburg district, Siberia.

No. 514. Ball of JASPER of a terra-cotta red, with creamy white spots. Diameter $1\frac{1}{2}$ ins. Cut, polished, and drilled at Oberstein.

Ekaterinburg district, Siberia.

No. 515. Ball of JASPER. Dull red, with green spots. Diameter 2 ins. Cut, polished, and drilled at Oberstein.

Ekaterinburg district, Siberia.

No. 516. Ball of SANDY BROWN JASPER. Diameter $1\frac{1}{2}$ ins. Cut, polished, and drilled at Oberstein.

Ekaterinburg district, Siberia.

No. 517. Ball of JASPER of a white ground. Striped black like a zebra. Diameter $1\frac{6}{8}$ ins. Cut, polished, and drilled at Oberstein.

Ekaterinburg district, Siberia.

No. 518. Ball of BROWN JASPER, with black and slate coloured markings. Diameter $1\frac{1}{2}$ ins. Cut, polished, and drilled at Oberstein.

Ekaterinburg district, Siberia.

No. 519. Ball of VARIEGATED JASPER A Chalcedonic base, with reddish-brown markings. Diameter $1\frac{1}{2}$ ins. Cut polished, and drilled at Oberstein.

Ekaterinburg, Siberia

No. 520. Ball of JASPER. A white base, with leek-green stripes. Diameter 1½ ins. Cut, polished, and drilled at Ober-stein.

Ekaterinburg, Siberia.

Porcelain Jasper.

No. 520A. PORCELANYTE or PORCELAIN JASPER. White. Really a naturally baked clay, having the fracture of flint, differing from true Jasper by being fusible before the blow-pipe. Formed by the baking of clay-beds when they consist largely of feldspar. Such clay-beds are sometimes baked to a distance of thirty or forty rods from a trap-dike, and over large surfaces by burning coal beds. Metamorphic. 5½ by 3¾ ins.

Bohemia.

QUARTZ ROCK OR QUARTZYTE.

No. 1. Granular Quartz.

No. 521. GRANULAR QUARTZ. Quartzyte, Granuline or Granulina, a variety of white granular pulverulent Silica encrusting lava. It is pure silica, from the scoriae of Etna. 4¼ by 3½ ins.

——— Mount Etna.

No. 2. Quartz Conglomerate.

No. 522. CONGLOMERATE OF QUARTZ consisting of fine large, distinct round and spheroidal pebbles of Silica, including Jasper and Chalcedony, some tinged a deep red, probably by iron oxide; others yellow, and of a violet tinge, some black, and some white as well as brown, naturally cemented together by siliceous cement, forming a most important and striking conglomerate. Extremely hard and difficult to cut. Very beautiful, and highly polished on one side. 10 by 7½ ins.

Harpenden, Berkshire.

A beautiful and remarkably characteristic specimen of Conglomerate unusually large, and consisting of particularly bright pebbles. A similar example to this is in the collection of Professor John Ruskin, at Sheffield.

No. 523. CONGLOMERATE of white and grey QUARTZ, with vivid red irregular-shaped specimens of Jasper, some looking as if in strata; associated with other brown and pale yellow pebbles of Silica, cemented together with a fine Quartzyte. 6¼ by 5 ins.

India.

A most striking and important conglomerate.

No. 524. QUARTZ CONGLOMERATE of rough pebbles, not well defined, of a bright red colour with patches of yellow Jasper, cemented naturally by Silica. An irregular boulder-shaped specimen. One side polished. 4¼ by 3 ins.

Louisa Creek, Wellington, Australia.

PSEUDOMORPHOUS QUARTZ.

(a) *Haytorite.*

No. 525. HAYTORITE,* so named from its locality, is a pseudomorphous variety of Quartz after Datholite, a boro-silicate of lime. 1⅛ by 1 in.

Haytor, Devonshire.

A rare and interesting pseudomorph.

---

b) *Babel Quartz.*

No. 526. BABEL QUARTZ, a white pseudomorphous Quartz, which has on the under surface impressions of cubes of Fluor Spar arising from its having been deposited over it. 3¼ by 1¼ ins.

Beer-Alston, Devonshire.

An interesting pseudomorph, called " Babel " Quartz through it presenting an appearance of towers, suggestive of the Tower of Babel.

* *C. Tripe, "Phil. Mag.," I., 40, 1827.*

Silicified Corals.

No form of Silica is more interesting than the siliceous corals and sponges. Many of the corals were composed of carbonate of lime, one of the softest materials, and were changed into one of the hardest forms of Silica (Hornstone), retaining their primary form and exhibiting clearly the canals and septa as well as their various orifices, fibres, and extremely delicate construction, although they have, by the gradual infiltration of Silica in solution, passed through this great pseudomorphous change, one of Nature's most marvellous processes. In many cases the formation is so distinct that there is no difficulty in classifying and allotting them approximately their proper position amongst the living species.

The West Indies, particularly Antigua, seems to have undergone some marvellous geological disturbance, whereby Silica in an incandescent or soluble state has spread itself over the whole country, producing by infiltration many thousands of specimens of silicified corals and sponges.

———

(c) *Silicified Corals.*

No. 527. SILICIFIED CORAL. Transverse section. Greyish-white, with patches of brown, exhibiting distinctly radiate structure and formation of the coral in an extremely hard variety of Silex. Polished well. 8 by 6 ins.

Antigua, West Indies.

No. 528. SILICIFIED CORAL. Quite translucent. The structure and septa are of a cloudy white with a brownish border. Complete section with outside bark showing original form. 5½ by 4 ins.

Antigua, West Indies.

No. 529. SILICIFIED CORAL. Genus Astræa, with structure and septa clearly exhibited. Brownish red nearly all over. 5¾ by 4 ins.

Van Diemen's Land.

No. 530. SILICIFIED CORAL. Structure and septa clearly defined, divided by bluish Chalcedony. Grey Hornstone and Chalcedony matrix. Polished on one side. Opaque. 4 by 3 ins.

Bandah, North West Province of India.

No. 531. SILICIFIED CORAL. Minute formation of the Genus Isastræa in grey Hornstone. Polished on one side. 2 by 2 ins.

Van Diemen's Land.

No. 532. SILICIFIED CORAL. Clearly exhibiting structure. Brown septa on a whitish ground. Shell-shaped. Well polished all over. 2½ by 1⅞ ins.

Bandah, North West Province of India.

No. 533. SILICIFIED CORAL. Pinkish or salmon hue to reddish. Exhibiting radiated formation very clearly, with minute crystalline Quartz often occurring in the septa. Thin section, beautifully polished on both sides and partly translucent. 3½ by 2¾ ins.

Cut and polished on the Marne, but from Ekaterinburg, Siberia.

No. 534. SILICIFIED CORAL. Pinkish or salmon to red. Thin section. Exhibits structure of Coral with radiation very clearly, with minute crystalline Quartz often occurring in the septa. Similar to No. 533. Beautifully polished on both sides and partly translucent. 3½ by 2⅜ ins.

Ekaterinburg, Siberia.

No. 535. SILICIFIED CORAL. Pinkish or salmon to red. Section exhibiting radiated structure of coral distinctly, with minute crystalline Quartz often occurring in the septa. Similar to Nos. 533 and 534. Beautifully polished on both sides and partly translucent. 3⅝ by 2¾ ins.

Ekaterinburg, Siberia.

No. 536. SILICIFIED CORAL SECTION. Salmon-pink to red. Exhibiting radiated structure clearly, with minute crystalline Quartz often occurring in the septa. Beautifully polished on both sides and partly translucent. 3½ by 2⅞ ins.

Ekaterinburg, Siberia.

No. 537. SILICIFIED SLAB, probably of a Madrepore. Buff colour with brown border. Of an indistinct mottled appearance. Polished on one side. 4 by 3½ ins.

Antigua, West Indies.

No. 538. CORAL, SILICIFIED. Structure well defined. Dark on a white ground. Oval. Polished all over. 1 by ¾ in.

Antigua, West Indies.

No. 539. CORAL, SILICIFIED. Structure well exhibited in brown septa on a whitish ground. Oval. Polished all over. 1 by ¾ in.

Antigua, West Indies.

No. 540. CORAL, SILICIFIED. Structure most clearly defined ; a brown centre, surrounded by minute chambers, on a whitish ground. Oval. Polished all over. 1 by ¾ in.

Antigua, West Indies.

No. 541. CORAL, SILICIFIED. On a greyish ground. Structure quite distinct. Oval. Polished all over. 1 by ¾ in.

Antigua, West Indies.

No. 542. CORAL, SILICIFIED. Large radiated structure clearly exhibited. Black lines on a white ground. Oval. Polished all over. 1 by ¾ in.

Antigua, West Indies.

No. 543. CORAL, SILICIFIED. Structure clearly defined ; white on a brownish red base. Oval. Polished on both sides. 1 by ¾ in.

Antigua, West Indies.

No. 544. CORAL, SILICIFIED. White on a grey ground. Structure visible under a magnifying glass. Oval. Polished on both sides. 1 by ¾ in.

Antigua, West Indies.

No. 545. CORAL, SILICIFIED. Grey. Structure plainly visible. Oval. Polished on both sides. 1 by ¾ in.

Antigua, West Indies.

No. 546. CORAL, SILICIFIED. White on a light brown ground. Structure plainly visible. Oval. Polished on both sides. 1 by ¾ in.

Antigua, West Indies.

No. 547. CORAL, SILICIFIED. Structure plainly visible in white on a grey ground. Oval. Polished on both sides. 1 by ¾ in.

Antigua, West Indies

No. 548. CORAL, SILICIFIED. A white concentric structure on a brown base. Oval. Polished on both sides. 1 by ¾ in.

Antigua, West Indies.

No. 549. CORAL, SILICIFIED. Exhibiting structure plainly. White on a grey ground. Oval. Polished all over. 1 by ¾ in.

Antigua, West Indies

No. 550. CORAL, SILICIFIED. Exhibiting structure in white and dark septa. Oval. Polished on both sides. 1 by ¾ in.

Antigua, West Indies.

No. 551. CORAL. Partly solidified, but still retains carbonate of lime, as it can be scratched with a knife. Structure visible on a brown base. Oval. Polished on both sides. 1 by ¾ in.

Antigua, West Indies.

No. 552. CORAL. Embedded in flint. White on a grey base. Oval. Well polished on both sides. 1 by ¾ inch.

Antigua, West Indies.

No. 553. CORAL, SILICIFIED. Globular in shape with thin black lines round the circles, on a brown base. Oval. Polished on both sides. 1 by ¾ in.

Antigua, West Indies.

No. 554. CORAL. White. Embedded in grey flint. Oval. Well polished all over. 1 by ¾ in.

Antigua, West Indies.

No. 555. CORAL, SILICIFIED. Plainly exhibiting structure. White and reddish. Oval. Well polished all over. 1 by ¾ in.

Antigua, West Indies.

No. 556. CORAL, SILICIFIED. Globular formation in concentric white circles. Oval. Polished all over. 1 by ¾ in.

Antigua, West Indies.

No. 557. CORAL, SILICIFIED, with fortification structure (partly Agate). Pinkish tint with white. Oval. Well polished on both sides. 1 by ¾ in.

Antigua, West Indies.

No. 558. CORAL, partly silicified and partly consisting of carbonate of lime. Yellow spots on a violet brown base. Oval. Polished on both sides. 1 by ¾ in.

Antigua, West Indies.

Pegmatyte.

Pegmatyte or Graphic Granite.

No. 559. PEGMATYTE or GRAPHIC GRANITE. Black-looking Quartz, arranged in a parallel position in whitish Feldspar. A large oval specimen, with bevelled edge, cut for a paper-weight. Well polished. 6¾ by 4⅛ ins.

Ekaterinburg, Siberia.

A most interesting and characteristic example, consisting of Quartz and Feldspar, the former arranged in cross fractures, presenting the appearance of written characters. Called also "Hebrew Stone."

No. 560. PEGMATYTE or GRAPHIC GRANITE, from γράφω= "I write." Small specimen of Quartz in a white Feldspar with a metallic reflection arranged in a parallel position. Polished. 2½ by 1¾ ins.

Ekaterinburg, Siberia.

M

OPAL.

OPAL is a hydrated Silica occurring in a different molecular state to Quartz, the hardness and specific gravity being less whilst it is hydrous, always containing some percentage of water, even up to 12 per cent. It is, undoubtedly, one of the most beautiful gems in existence, although looked upon, even in these enlightened days, by many with superstition, particularly by the French, whose invariable phrase upon the mention of its name is " Cela porte malheur," perhaps originating through such an episode as that of the Baroness Hermione of Arnheim in Sir Walter Scott's novel *Anne of Geierstein*. It derives its name from the Latin *Opalus* or *Opalum ;* but Pliny describes it under the name of *Pæderos*, or " Boy's Love," and refers to the extraordinary beauty of the Precious or Noble Opal reflecting all the prismatic colours in refulgent tints, now exhibiting one hue with indescribable brilliancy, and now another. Pliny says it is " made up of the glories of the most precious gems. To describe it is a matter of inexpressible difficulty ; there is in it the gentler fire of the ruby; there is the brilliant purple of the amethyst; there is the sea-green of the emerald, all shining together in an incredible union. Some aim at rivalling in lustre the brightest azure (armenium) of the painter's pallet, others the flame of burning sulphur, or of a fire quickened by oil."

Its wonderful colour has always been attributed to the presence of microscopic cavities, which theory was originated by Sir D. Brewster ; [*] but, according to Behrends, it owes its

* " *Edin. Phil. Journ.*," 38, 385, 1845.

beautiful colours to "thin curved lamellæ of opal,* whose refractive power differs slightly from the mass by 0·1." These lamellæ, Dana says, "are conceived to have been formed originally in a parallel position, but have been changed, bent, and finally cracked and broken in the solidification of the ground mass."

Precious Opals of very fine quality are not often large The celebrated Kaschau opal, valued at £3,000, was only the size of half-a-crown, whilst the stone, set as a ring, in the possession of Nonius, a senator who, Pliny tells us, was proscribed because Antonius coveted it, was not larger than a nut, although valued at 2,000,000 sesterces, equal to about £17,000 of our money. On being thus proscribed he took to flight, carrying with him of all his wealth nothing but this ring. One extraordinary specimen is in the Cabinet of Vienna, and weighs 17 ozs., but it is not quite detached from the matrix, and unfortunately is full of fissures. Amongst the French Crown jewels were two wonderful opals, one mounted as a clasp to the Imperial cloak, whilst the ex-Empress Eugenie possessed another very fine stone, to which, it is said, she attributed her great misfortune. This specimen came subsequently into the possession of the author, who disposed of it to the late John Noble, Esq., of Park Place, Henley-on-Thames.

The finest precious opals are found in a cellular porphyritic rock at Czerwenitza, near Kaschau, Hungary. When the prismatic colours present a variegated play, and are distributed regularly over the surface, it is termed the Harlequin Opal, on account of it somewhat resembling the motley tints of the Harlequin's dress. This variety, when *very fine*, is greatly esteemed, but if only of second-rate beauty, the large broad and more distinct flashes are preferred. Other fine specimens come from Mexico, whilst of late years magnificent specimens in a red ironstone Jasper have been discovered on the Barcoo river, and at other localities in Queensland.

* "*Ber. Ak. Wien.*," 64 (1) 1871.

The common opals are found also in Hungary, Bohemia, Saxony, and other localities.

OPAL (Latin, *Opalus* or *Opalum*). *Pæderos* of Pliny Quartz resinite. Amorphous, massive, sometimes small reniform, stalactitic or tuberose; also earthy. Hardness 5·5 to 6·5. Specific gravity 1·9 to 2·3; when pure 2·1 to 2·2. Lustre vitreous; frequently sub-vitreous. Colour white, yellow, red, brown, grey, blue; generally pale. Dark coloured or so-called Black Opals take their hue from foreign admixtures. Occurs seldom larger than a hazel nut. Possesses often rich play of colours or different colours, by refracted and reflected light. Streak white. Composition, silica-like Quartz, with a varying amount of water, SiO_2nH_2O. M. Damour of Paris made the analysis of a Precious Opal from Hungary, which yielded:

Silica	93·90
Water	6·10
	100·00

The water is sometimes regarded as non-essential, but as a rule is present from 3 to 9 and sometimes 12 per cent. Many impurities are often associated with it, including ferric oxide, alumina, lime, magnesia, and alkalies.

Opal.

No. 1. *Precious Opal.*

No. 561. PRECIOUS OPAL. Noble Opal. *Pæderos* of Pliny [*] Mentioned also as *Sangenon* or *Tenites*. Derived from the Greek word ὤψ, signifying the eye, the ancients believing it possessed the power of strengthening the eye. An analysis of Noble Opal with iridescent reflections by Klaproth yielded:

Silica ...	... 90·00
Water...	... 10·00
	100·00

Thin section of blue and green, with a little red, in refulgent tints, embedded in porphyry, attached to a base of slate. Rectangular, with cut corners. $3\frac{3}{4}$ by $1\frac{3}{4}$ ins.

Czerwenitza, near Kaschau, Hungary.

[*] *Lib.* xxxvii., *cap.* 21, 22.

No. 562. PRECIOUS OPAL, principally of a bluish tinge in effulgent tints disseminated over a matrix of red Ironstone Jasper of a banded nature. $4\frac{1}{4}$ by 3 ins.

Barcoo River, Queensland.

No. 563. PRECIOUS OPAL. *Pæderos*, Pliny. Blue and green in effulgent tints, embedded in red Ironstone Jasper. Also in veins with a liver-coloured, opaque Jasper. 2 by $1\frac{3}{4}$ ins.

Barcoo River, Queensland.

No. 564. PRECIOUS OPAL. Blue and green, disseminated through a brown and red Ironstone Jasper. $1\frac{3}{8}$ by $1\frac{1}{2}$ ins.

Barcoo River, Queensland.

No. 2. Fire Opal.

No. 565. FIRE OPAL. *Feuer-Opal.* Of a hyacinth-red, with fire-like reflections, becoming irised on turning. An analysis made of a Fire Opal from Zimapan, Mexico, by Klaproth yielded—

Silica ...	92·00
Ferric Oxide	0·25
Water ...	7·75
	100·00

Translucent. $1\frac{1}{2}$ by $1\frac{1}{2}$ ins.　　　　Zimapan, Mexico.

No. 3. Common Opal.

No. 566. COMMON OPAL, of a yellow-green colour, with a resinous lustre and partly translucent. Polished on one side. $2\frac{1}{2}$ by $1\frac{3}{4}$ ins. An analysis yielded :

Silica	82·75
Ferric Oxide ...	3·00
Alumina	3·50
Lime...	0·25
Water	10·00
	90·50

Czerwenitza, Hungary.

No. 567. COMMON OPAL, green colour, with yellow associated with a botryoidal silica, small. Partly translucent. $2\frac{1}{2}$ by 2 ins.

Kaschau, Hungary.

(a) *Wax Opal.*

No. 568. WAX OPAL or RESIN OPAL, of a honey-yellow, with a resinous lustre. Transparent. $2\frac{3}{8}$ by $2\frac{1}{8}$ ins.

Kaschau, Hungary.

No. 569. CHALCEDONIC WAX OPAL or RESIN OPAL, partly Chalcedony and partly Opal, of a yellowish green. Resinous lustre. Translucent. Polished on one side. $2\frac{1}{2}$ by $2\frac{1}{4}$ ins.

—— Kaschau, Hungary.

No. 4. Rose Opal.

No. 570. ROSE OPAL, or *Quincite* of Berthier, so named from its locality, *Quincy.* Carmine-red in a deposit of limestone. Hydrated silica, with magnesia and iron. The colour is attributed to organic matter. Unpolished specimen. $2\frac{3}{8}$ by $1\frac{3}{4}$ ins. An analysis by Rose yielded :

Silica	...	...	...	54·00
Water	...	...	...	17·00
Magnesia	...	...	...	19·00
Ferrous Oxide	...	...		8·00
				98·00

—— Quincy, France.

No. 5. Prase Opal.

No. 571. PRASE OPAL, of a beautiful Chrysoprase green colour. A comparatively new discovery. $2\frac{3}{4}$ by $1\frac{3}{4}$ ins.

Baumgarten, Frankenstein, Silesia.

No. 6. Wood Opal.

No. 572. LITHOXYLE or WOOD OPAL. Holz-opal *Germ.* Wood petrified by Opal, and coloured by hydrated peroxide of iron. Exhibits woody structure with parallel lines. Polished surface. $2\frac{3}{4}$ by $1\frac{5}{8}$ ins. An analysis of Wood Opal yielded :

Silica	...		...	...	93·00
Ferric Oxide		...	...	0·38
Alumina	...		...	...	0·13
Water	...		...	...	6·13
				99·64

Near Hobart Town, Tasmania.

No. 7. Fiorite.

No. 573. FIORITE (two specimens). Pearl Sinter. White stalactitic and botryoidal pearly concretions detached from Tufa.

About 1 inch long. In glass tube. An analysis yielded:

Silica	...	...	...	...	87·67
Ferric Oxide and Alumina			...	0·71
Lime...	...	...	...	...	0·40
Water	...	...	...	...	10·40
Soda and traces of Potash			...	0·82
					100·00

Santa Fiora, Grossetta, Tuscany, Italy.

Derived from the decomposition of siliceous minerals in volcanic rocks, about fumaroles ; and deposited from the siliceous waters of hot springs, in part by the action of vegetation. At times it passes into a transparent glassy variety of opal called Hyalite.

No. 8. Geyserite.

No. 574. GEYSERITE. A white and grey porous concretionary mass, with a hardness of 5·0. 3 by 1½ ins. An analysis yielded Forchhammer :

Silica	...	...	...	...	84·43
Ferric Oxide	...	...	...	1·91
Alumina	...	...	...	...	3·07
Lime...	...	...	...	...	0·70
Water	...	...	...	...	7·88
Soda and trace of Potash...		...	0·92
Magnesia	...	...	...	...	1·06
					99·97

Bombiana, Bologna, Italy.

No. 9. Float Stone, Quartz Nectique.

No. 575. FLOAT STONE or QUARTZ NECTIQUE of Hauy, called also Schwimmstein, *Germ.* A light concretionary grey circular spongy mass, generally cavernous, which floats, through its spongy texture, on water. 2½ by 2 ins. An analysis yielded :

Silica	...	...	...	85·90
Alumina	...	...	...	0·70
Carbonate of Lime ...		...	9·10
Water	...	...	...	3·30
				99·00

Menil-Montant, near Paris.

Pliny describes a stone of *Scyros*, saying, " In the Isle of Scyros there is a stone, they say, which floats upon water when whole, but which falls to the bottom when broken into fragments," which is most probably floating Quartz.

FLUORIDES.

Fluorite or Fluor Spar Group.

THIS beautiful mineral is a Fluate of Lime or a Fluoride of Calcium, deriving its name from *fluo*—" to flow."

It occurs generally associated with metalliferous veins, particularly lead, and is used extensively as a flux, Nature providing the metal for our use, with the accompanying agent for its reduction—Fluor Spar. It was known to the ancients, and it has been lately suggested that the Murrhine or Myrrhine Vases of Pliny were really made of Fluor Spar, the description certainly agreeing with this mineral.

The material of which these Vases were made has been a disputed point with modern authorities—some inclining to the opinion that they were simply glass, others variegated Agate or Chalcedony, whilst the majority that they were Chinese porcelain. Flour Spar seems to answer to *all* the descriptions of the several classic authors, more particularly that of their colour ranging from purple to white, and then shading off to red. This material is improved greatly by being heated, which is not the case with agates, and therefore recalls the passage " *Murrheaque in Parthis pocula cocta focis.*"

The Murrhine Vases and Cups were brought by Pompey the Great to Rome after his Parthian Expedition, and Murrhine was described as " fit for making dishes and drinking-cups." This is quite in character with Fluor Spar, which is most suitable, and for the manufacture of which it is in modern times extensively used. As it is also stated that the Murrhine Vases were *first* introduced by Pompey, this could not allude to Agate, which was well known long *before* his

time. Another remarkable corroboration is that Fluor Spar
has been discovered in the neighbourhood of the Caspian
Sea—the locality of the Parthian Expedition. It is therefore only natural to suppose that the Murrhine Vases were
probably Fluor Spar, and not Agate. Another point is that
few fragments of the Murrhine Vases have been found,
which indicates not only their rarity, but the probability that
they were made of a soft material which would easily get
broken and crushed, like Fluor Spar, during the great lapse
of time, whereas had they been of Agate or Chalcedony, the
harder materials, no doubt many more fragments would have
been preserved; and the description of colour agreeing so
perfectly with the Fluor Spar, tends greatly to advance the
opinion that it was the mineral of which the vases were
made.

Specimens of a pale green variety of Fluor Spar, with a
border of Hornstone set in an ornament, was discovered at
an excavation in Rome, and was supposed to be the Murrhina.

Fluor Spar is found at many localities in Derbyshire, Cumberland, Devonshire, Cornwall, Saxony, Switzerland, etc., but
it is only to the Derbyshire variety that the name (a local
one) "Blue John," is given. It is made into vases, tazze,
and bowls, the finest specimens being found at Castleton,
which furnished years ago broad veins of deep and decided
colours cut from a part of the mine called the "Bull-beef"
vein. The mine has been exhausted for about half a century,
so that ornaments cut from this particular vein are greatly
sought after, and are getting rarer every day.

The finest specimens are cut from a nodular variety, the
colours of which are very fine, running in distinct zones or
bands. See Nos. 577, 578, 579.

The colours are attributed by Wyrouboff* to compounds
of carbon and hydrogen derived from a slight infusion of
organic matters in the solvent waters. He states that on
being subjected to heat the blue and violet change to purple,

* *"Bull. Soc. Ch,"* 5, 334, 1866.

and supposes that two CH substances, a blue and a red were present, the former more volatile and therefore leaving the colour reddish after partial heating.

The colours are often deepened by heat, particularly those of the amethystine hue. Great care has, however, to be shown, as it is fragile and easily cracked.

A variety of Fluor Spar called *Chlorophane* is noted for its beautiful phosphorescence, generally of a green hue. It turns to white if placed on a heated shovel, emitting in a dark room a most intense and beautiful emerald-green light. Etching on glass by means of Fluor Spar was practised as early as 1670 at Nuremberg.

Fluor Spar Section.

Fluorite or Fluor Spar—from the Latin, fluo "to flow"—being used as a flux to Galena and other ores.

Agricola says : *Fluores lapides gemmarum similis sed minus duri—qui ignis calore liquescunt* (whence he derives the name) *colores varii, jucundi,* (1) *rubri,* (2) *purpurei (vulgo amethysti),* (3) *candidi,* (4) *lutei,* (5) *cineracei,* (6) *subnigri, etc.* Berm. 458, 1529. Fluate of Lime. Derbyshire spar ("Blue John"). Chaux Fluatée, *Fr.* Flussspath, *Germ.* Isometric or cubical. Occurs in twins. Cleavage, octahedron perfect. Fracture sub-conchoidal. Brittle.

Hardness 4·0. Specific gravity 3·01 to 3·25

Lustre vitreous. Colour white, yellow, green, rose-red, violet-blue, sky-blue, violet-yellow, greenish-blue, violet-blue (common) ; red (rare). The colours are sometimes different, according as they are seen by reflected or transmitted light. Streak white. Transparent, sub-translucent. Exhibits bluish fluorescence and phosphorescence when heated slowly.

Comp. CaF_2.

Fluorine	48·9
Calcium	51·1
	100·0

Chlorine is sometimes present in minute quantities.

No. 1. Fluorite or Fluor Spar.

No. 576. FLUORITE or FLUOR SPAR. A fibrous variety called Derbyshire Spar, locally "Blue John." Magnificent large section of a deep bluish-violet, in zigzag and concentric formation. Blackish, opaque in centre, surrounded by bluish-violet, with exterior of a yellowish tinge ; has been treated artificially by being subjected to heat. Iridescent in some parts. Beautiful by transmitted light. Polished on both sides. 10 by 9½ ins.

Castleton, Derbyshire.

No. 577. FLUORITE or FLUOR SPAR. Fibrous variety called Derbyshire Spar, or, locally, "Blue John." Of a most beautiful dark purple and amethystine centre, with white bands, tinged with oxide of iron, and a white rim. Procured from the celebrated vein called the "Bull-beef" vein. Thick section. Greatly cracked. Has been subjected to heat. 7¼ by 6¼ ins. by ¾ in. thick.

Castleton, Derbyshire.

No. 578. FLUORITE or FLUOR SPAR. A fibrous variety called Derbyshire Spar. A beautiful dark purple and amethystine centre, with white bands tinged with oxide of iron. Greatly cracked all over. Thick section, cut from the same specimen as No. 577. 7¼ by 6¼ ins.

Castleton, Derbyshire.

No. 579. FLUORITE or FLUOR SPAR, Derbyshire Spar, "Blue John." Section of a violet-blue with a black centre, fortification layers, surrounded by a reddish coloured border. Artificially deepened in colour by heat. From the "Bull-beef" vein. Polished. 6¼ by 6 ins.

Castleton, Derbyshire.

No. 580. FLOURITE or FLUOR SPAR. "Blue John." Section. A translucent, white, semi-crystalline centre tipped with violet, a deep band of blackish-blue surrounded by light blue fortification lines and semi-crystalline white fluor with a violet edge. All deepened by heat. This and the colours of the four following specimens are described by transmitted light. Polished on both sides. 4¾ by 3¾ ins.

Castleton, Derbyshire.

A most exquisite specimen by transmitted light.

No. 581. FLOURITE or DERBYSHIRE SPAR. " Blue John."
Section. White and violet-blue semi-crystalline centre, with
a deep band of dark blue encircled by amber coloured Fluor
Spar. Well polished on both sides. Exhibits outside edge
all round. Most beautiful by transmitted light. 4 by $3\frac{3}{4}$ ins.

Castleton, Derbyshire.

No. 582. FLUORITE OR FLUOR SPAR. Fibrous section.
Centre of a lovely dark, nearly black-violet or plum colour,
paling off to a lighter hue, encircled by a broad layer of white,
running into a lovely light violet hue in fortification, with
another band deeper but of the same colour with whitish
exterior, produced probably by heat. Transparent and
exquisite by transmitted light. 4 by $3\frac{1}{2}$ ins.

Castleton, Derbyshire.

No. 583. FLUORITE. Derbyshire Spar. Fibrous. Section
consisting of an amethystine purple centre with band of white
tinged red, produced probably by heat, with light violet and
an amber coloured Fluor on the exterior. 5 by $3\frac{1}{4}$ ins. Described by transmitted light.

Castleton, Derbyshire.

No. 584. FLOURITE or FLUOR SPAR. A large specimen
in octahedral crystals of a light apple-green colour upon
Hornstone. The crystals are well developed and show the
octahedral form distinctly. $9\frac{1}{2}$ by 6 ins.

Puy de Dôme, Auvergne.

No. 585. FLUORITE or FLUOR SPAR. A group of beautiful
green cubical crystals, quite translucent. $7\frac{1}{2}$ by $5\frac{1}{2}$ ins.

Alston Moor District, Cumberland.

A most lovely coloured specimen of pure limpid Fluorine.

586. FLUOR SPAR. Section of a light green colour in
irregular semi-crystalline bands, associated with white Fluor
Spar. Polished on one side. 6 by $3\frac{3}{4}$ ins.

Hartz Mountains, Germany.

CARBONATES.

A. ANHYDROUS CARBONATES.

Calcite Group.

THE Calcite Group (from the Latin *Calx* "lime") or Carbonate of Lime covers a wide range of minerals, from the purest Iceland Spar through the whole range of limestones, marble, and chalk ; the marbles themselves, of endless variety, used for architectural purposes, constituting a class upon which many books have been written. Calcite constitutes the whole of the chalk formation and occurs in nearly all others. Its crystallization is marvellous, hundreds of secondary forms of the rhombohedron being known, all the angles of which have been taken and described by the learned and erudite crystallographer, Professor Descloiseaux, of the Jardin des Plantes of Paris. It was used in its form of marble largely by the ancients for sculpture, and Pliny devotes a good deal of space to it. The finest statuary marble is the fine-grained "Carrara" (specific gravity 2˙71 to 2˙72), from the quarries of Monte Crestola and Monte Sagro, which have been worked for nearly 2,000 years in Italy, it being fine-grained and firm in texture, as well as of a beautiful white. The most celebrated antique marble was the Parian, the Lychnites of the ancients, which is characterized by a lamellar structure and very small semi-translucent flakes, with a specific gravity of 2˙70 to 2˙71, from the Isle of Paros, the birthplace of the great Phidias and Praxiteles, and of which the most lovely and historical statues, such as the Venus de Medici, the Venus of the Capitol, and the Venus of Milo are made, whilst the Pentelicon Marble, with a gravity of 2˙716, and those from the quarries of Mont Hymette, near Athens, with a lamellar grain also, although not so white in colour, being slightly grey, whilst the *Luni* marbles of Tuscany are the next in quality. Architectural marbles are of every colour. The *Cipolin* of Italy is white with pale green reflections from Talc ; the *Giallo Antico* is ochre and cream-yellow with white spots ; the Siena or *Brocatello de Siena (Marmor Numidicum)* is yellow, veined with bluish-red and sometimes purple, with a gravity of 2·67 to 2·70 ; the *Mandelato* is light red with yellowish-white spots ; the *Bardiglio* is grey clouded with serpentine from Corsica ; the *Nero-antico* is an ancient deep black marble ; the *Paragoni* is a modern one of a fine black colour from Bergamo ; the *Panno-di-Morti* is another black marble with a few white fossil shells; the *Rosso-antico* is a red marble ; whilst a greyish blue-coloured marble veined with white is called *Turquoise-blue* marble, and Verd-antique is the clouded green-coloured variety containing serpentine. Scores of other varieties are known under various names taken from their distinctive colours or localities. The Lumachelle or Fire Marble is rare, and is well represented by Nos. 597 to 599. Ruin Marble is of a yellow to brown colour, which when cut perpendicularly to the planes of the layers, and polished, exhibits figures bearing some resemblance to ruins, fortifications, temples, etc., due to the infiltration of iron-oxide. A very fine specimen is in the collection—No. 601. The limestone rocks of Devonshire, black, white, pink, and other colours, largely used for building and other purposes, enclose fossil corals which exhibit their structure and septa very well, as will be seen by the fossil corals, Nos. 605 to 627

Calcite occurs greatly in stalactites as well as stalagmites, forming the floors and *walls of caverns.*

Varieties based chiefly upon Crystallization and accidental impurities.

CALCITE. *Marmor* Pliny : Lapis Calcarius. Calc Spar. Carbonate of lime. Rhombohedral. Fracture conchoidal obtained with difficulty. Hardness 3·0, but varying with the direction on the cleavage face. Specific gravity 2·713

to 2'714. Colour white, or colourless, also grey, red, green, blue, violet, and yellow. Transparent to opaque.
Composition—Carbonate of lime.

Carbon Dioxide	...	... 44'0
Lime ...	... ...	... 56'0
		100'0

Small quantities of magnesium, iron, manganese, zinc, and lead may be present, replacing the calcium.

No. 1. Calcite—Iceland Spar.

No. 587. ICELAND SPAR or double refracting spar. *Doppelspath* of the Germans. Calcite. Carbonate of lime. Large rhomb produced by cleavage. Pure transparent variety of calcite ; double refraction very strong, separating widely the two rays into which the ray of incidence is divided, producing a double figure or image of a spot or line through a cleavage fragment.

An analysis from this locality yielded Biot and Thenard:

Carbon Dioxide	...	...	... 43'045
Lime	...	...	... 56'327
Water	...	...	... 0'628
			100'000

White, beautifully iridescent exhibiting the prismatic colours of the rainbow through fracture. Hardness 3'0. Specific gravity 2'713 ; in pure crystals 2'723 Beudant. 6 by 5 ins.
Helgustadir on the Eskefiord, Iceland.

This interesting transparant and purest of carbonates of lime is found in a very small mine, being in fact only a large cavity in basalt, and has been worked for a considerable number of years, under the supervision of the Danish Government—latterly, however, very sparingly, as the supply is becoming exhausted. At the present the mine is entirely closed by the direction of the Danish Parliament.

It is a most valuable mineral and extensively used for making polarizing prisms for microscopes and other optical intruments. Magnificent transparent and also crystallized specimens were brought twenty years ago from the mine by Herr Carl Franz Siemsen of Hamburg. A very fine macle, or twin crystal, is exhibited in the British Museum at South Kensington, cleaved by the author from an immense block nearly two feet high. In later years Herr Tulinius, a Dane, brought it to Europe in large quantities, but the supply for the present is entirely stopped. It crystallizes well, and is often associated with Stilbite, one of the Zeolitic group, occurring in tufts.

No. 588. CALCITE or carbonate of lime, containing a mass of brilliant needle-like crystals of brass-yellow Pyrite, iron pyrites, the sulphide of iron. Cut and polished on one side. 2½ by 2½ ins.

————— Matlock, Derbyshire.

No. 2. Fibrous and Lamellar Variety.

No. 589. SATIN SPAR. In irregular parallel strata, well defined. Formerly was placed with Aragonite but now it appears to be purely a fibrous calcite, its specific gravity being 2·720 or, according to Damour, 2·727, with a hardness a little less than Aragonite. White, grey, and of a pinkish hue. Large important example. 7 by 5¼ ins.

Ness Rocks, Shaldon, near Teignmouth.

No. 590. SATIN SPAR. Faserkalk. Atlasspath, *Germ.* A beautiful variety of pure white calcite in fibres of a silky-lustre, which it displays to great advantage when polished, and from which it takes its name of Satin Spar. Polished. 2 by 1⅞ ins.

Found in a cliff of decomposing shale in thin veins by the author, at Alston Moor.

Cumberland.

A fibrous gypsum (sulphate of lime) is sometimes called Satin Spar, but this true variety is much harder and effervesces with acid. It often contains 4·25 per cent. of carbonate of protoxide of manganese, which communicates to it a rose tinge

No. 591. SATIN SPAR. Faserkalk, *Germ.* Pure white. Fibrous, with a silky lustre. Polished. 1⅞ ins. square.

Alston, Cumberland.

—————

No. 3. Granular Massive to Crypto-crystalline varieties.

(a) *Mexican Onyx.*

No. 592. CALCITE. Carbonate of lime. White variety called "Mexican Onyx." Stalagmitic; from the floors of caverns, being derived from depositions from the roofs or walls; in irregular curved layers in Agate or Onyx-like bands. Translucent to transparent. Polished on both sides. 7½ by 3½ ins.

Tecali, Puebla, Mexico.

This is the variety so largely used for decorative purposes, such as table-tops, vases, etc. Being soft, it is easily manipulated and takes a high polish. The word Onyx is a misnomer, and should not have been used, as the Onyx is Silica, and not Lime.

No. 593. CALCITE. Variety called " Mexican Onyx."
Stalagmitic. White, in beautiful onyx-like layers of great
delicacy. Translucent. Round. Polished on one side.
4⅜ ins. in diameter.

Tecali, Puebla, Mexico.

No. 594. CALCITE. Carbonate of lime. Variety called
" Mexican Onyx." White-clouded, with yellow markings.
Stalagmitic. Polished. 4⅞ by 3 ins.

Tecali, Puebla, Mexico.

____

(b) *Calc Tufa.*

No. 595. TUFA, Travertine, the *Lapis Tiburtinus* of Vitru-
vius and Pliny. Calc Tufa. A stalagmitic variety of carbonate
of lime, deposited from springs or a river, sometimes scores
of feet in thickness. The word *Tophus* is used also by Pliny,
from which it derives its name. Mammillated, and looking
like fortification Agate. Deep reddish-brown, white and
brown in half circles distinctly showing aqueous deposition.
Polished. 14 by 8⅜ ins.

California.

A most useful variety of Calcite, largely used as a foundation for
ferneries, etc.

No. 596. TUFA. Calc Tufa. Tophus of Pliny. Octagonal.
Red mammillated specimen. Exhibiting aqueous formation.
Well polished. 3 by 2½ ins.

California.

____

(c) *Lumachelle Marble.*

No. 597. LUMACHELLE or FIRE MARBLE. A very dark
brown compact shell-marble with brilliant flammeate or
chatoyant iridescent reflections proceeding from the embedded remains of Nautili and other shells. Specific gravity
by Damour 2·746. 6 by 3½ ins.

Bleiberg, Carinthia.

A most beautiful, valuable, and well known variety of marble used
for inlaying the tops of snuff-boxes, etc., taking a very high
polish.

N

No. 598. LUMACHELLE or FIRE MARBLE. A dark brown compact shell-marble with brilliant chatoyant iridescent fire-like reflections proceeding from the remains of embedded shells. Polished on one side. $4\frac{1}{2}$ by $8\frac{1}{4}$ ins.

Bleiberg, Carinthia.

No. 599. LUMACHELLE or FIRE MARBLE. Very dark brown compact limestone with brilliant flammeate and iridescent green and blue reflections proceeding from the remains of embedded shells. Polished on one side. 5 by $3\frac{5}{8}$ ins.

Bleiberg, Carinthia.

(d) *Laminated Oolite.*

No. 600. LAMINATED OOLITE, Calcaire Oolitique (Rogenstein). A slab of granular limestone in minute globular concretions, cemented by a calcareous cement. Buff colour, with black delicate dendritic or tree-like markings of Pyrolusite, the dioxide of manganese. Remarkable arborescent appearance. 7 by $4\frac{3}{4}$ ins.

Pappenheim.

(e) *Ruin Marble.*

No. 601. Magnificent specimen of RUIN MARBLE, *Marble ruiniforme*, the *Ludas Helmontii* of the ancient mineralogists. A compact calcareous marl of a buff colour, exhibiting a resemblance to fortifications, ruins, landscapes, etc., due to the infiltration of oxide of iron through the layers of marble, which are then cut perpendicularly to the planes of the strata. This specimen represents somewhat the ruins of a castle next to water, which, being as it were, transparent, gives a view of rocks beneath it. In a semi-polished state. $8\frac{1}{2}$ by $8\frac{1}{4}$ inches.

Banks of the Arno, Tuscany.

No. 602. RUIN MARBLE, *Ludus Helmontii* of the ancient mineralogists. Yellow with a portion of the interior of a slaty-bluish tint. Polished on one side. $3\frac{3}{8}$ by $3\frac{1}{4}$ ins.

Siena, Italy.

No. 603. CALCITE, or carbonate of lime, coloured pink, probably by a trace of cobalt. Polished on one side. $4\frac{3}{4}$ by $2\frac{1}{2}$ ins.

Puebla, Mexico.

(f) *Limestone in which all Devonshire Marbles and Devonian Corals are included.*

No. 604. CALCITE (carbonate of lime), tinged red with iron oxide. Locally called " Babbacombe bloodstone," but not in any way a Silica. Polished. 3 by 1¾ ins.

Babbacombe, Devonshire.

No. 605. Conglomerate rock (pebble) of carbonate of lime, containing *Connopora* and sponge. Black, base banded with yellow layers. Well polished on one side. 3½ by 2½ ins.

Ness Rocks, Devonshire.

No. 606. FOSSIL CORAL (*Amphipora ramosa*) in limestone. Structure plainly visible. Polished one side. 3½ by 2 ins.

Ness Rocks, Devonshire.

No. 607. FOSSIL CORAL (*Favosites sp.*) in Devonian limestone. Red, tinged probably with iron, striated and black. Exhibits structure of the coral distinctly. Polished on one side. 7¾ by 4½ ins.

Watcombe, Torquay.

No. 608. FOSSIL CORAL (*Cyathophyllum helianthoides*). Devonian limestone. Red and yellow. Structure plainly exhibited. Polished on one side. 6¼ by 5½ ins.

Cockington, Torquay.

No. 609. FOSSIL in limestone (*Stromatopora digitatum*). Greyish, in irregular circles. Slightly stained with iron. Polished on one side. 9⅛ by 5¼ ins.

Near Happaway, Torquay.

A large and important specimen.

No. 610. FOSSIL CORAL, in Devonian manganiferous limestone (*Acervularia pentagona*). The structure is most plainly exhibited. Pinkish hue. Well polished. 6½ by 3½ ins.

Ogwell, Devon.

No. 611. FOSSIL CORAL, pink, in manganiferous Devonian limestone (*Acervularia pentagona*). The structure is most plainly exhibited. Well polished.   3½ by 2¾ ins.

Ogwell, Devon.

No. 612.   FOSSIL CORAL (*Lithostrotion junceum*, Flem.). Structure beautifully distinct.   On a black ground of limestone. Well polished.   4⅜ by 2¾ ins

Clifton, Bristol.

No. 613. FOSSIL CORAL in Devonian limestone (*Favosites cervicornis*, Blainville). Whitish coral. A thick specimen. 6½ by 4 ins.

Bishop's Teignton, Teignmouth.

No. 614. FOSSIL CORAL (*Favosites cervicornis*, Blainville) in Devonian limestone.   Thin section.   6½ by 4 ins.

Bishop's Teignton, Teignmouth.

No. 615. FOSSIL CORAL in Devonian limestone (*Favosites cervicornis*, Blainville).   Well polished.   4 by 3 ins.

Bishop's Teignton, Teignmouth.

No. 616. FOSSIL CORAL in Devonian limestone, allied to *Cyathophyllum cæspitosum*, Goldf.   White septa are exhibited on a black ground.   4½ by 2½ ins.

Ness Rocks, Shaldon, Devon.

No. 617. FOSSIL CORALS in Devonian limestone.   Structure discernible, but of a mottled appearance.   Well polished. 3⅝ by 2¾ ins.

Parson and Clerk Rock, Teignmouth, Devon.

No. 618. FOSSIL CORAL (*Cyathophyllum cæspitosum*, Goldf.) in Devonian limestone, containing also Connopora.   Whitish upon a purple ground.   Well polished.   3 by 2¼ ins.

Teignmouth, Devon.

No. 619. Fossil Coral. White on grey ground. Locally called the "Globe Stone." Structure visible by the aid of a magnifying glass. Polished on one side. 5½ by 3⅛ ins.

Babbacombe, Devon.

No. 620. Fossil Shell, magnificent specimen of *Orthoceras Devoniensis*, exhibiting the chambers and internal as well as external formation very distinctly, embedded in veined limestone. An abnormally large and quite unique specimen of a fine flammeate red, probably caused by ferric oxide. In some parts it is black, and in others white, both aiding to exhibit the structure of the shell. 16¼ by 10½ ins.

Teignmouth, Devonshire.

No. 621. Fossil Coral (*Lithostrotion irregulare*, M'Coy). Grey septa in brownish, in Devonian limestone. 3¾ by 3⅛ ins.

Clifton, near Bristol.

No. 622. Fossil Coral (*Stromatopora Concentrica*) in Devonian limestone. A remarkable zigzag formation in thin white lines on a grey ground. Well polished. 5½ by 3¾ ins.

Parson and Clerk Rock, Teignmouth.

No. 623. Fossil Coral (*Lithostrotion basaltiforme*) in Devonian limestone. Structure and hexagonal formation plainly exhibited. Grey. Well polished. 3¾ by 3 ins.

Silverdale, Yorkshire.

No. 624. Fossil Coral (*Alveolites porosa*) in Devonian limestone. Of a fine Jaspery red colour, exhibiting structure well by the aid of a magnifying glass. Well polished. 5¾ by 4¼ ins.

Torre Abbey, Torquay.

No.625. FOSSIL CORAL (*Heliolites porosa*, Goldf.) in Devonian limestone. Irregular reddish bands. Structure minute but visible by aid of magnifying glass. Well polished. 3¼ by 2¼ ins.

Wolborough Rocks.

No. 626. FOSSIL CORAL (*Favosites cæspitosa*) in Devonian limestone. White septa on a black ground. Well polished. 5 by 4 ins.

Teignmouth, Devon.

No. 627. FOSSIL CORAL (*Stromatopora*) in Devonian limestone. Reddish and whitish tint. Well polished. 4¾ by 4 ins.

Teignmouth, Devon.

MALACHITE.

Basic Carbonates.

Few minerals are so well known or so attractive as Malachite, the Carbonate of Copper, in consequence of its distinctive green colour, so refreshing to the eyes, even if crude. The different shades are arranged in concentric strata, marking the successive deposition of the mineral percolating through copper-bearing rocks, and depositing the carbonate in fissures, which upon the evaporation of the water forms a series of layers similar to the manner in which stalagmites and stalactites of carbonate of lime are formed in Derbyshire and elsewhere by the percolation of water through the limestone. Speaking of a large mass of Malachite exhibited in the Great Exhibition of 1851, Sir Roderick Murchison graphically describes the formation as follows:

"The geological interest attached to this mass lies in the indication it affords that the substance called Malachite has been formed by a *cupriferous solution*, which has successively deposited its residue in a stalagmitic form. *Mutatis mutandis*, this mass has only to be viewed as formed of calcareous spar (carbonate of lime), and it presents every one of the features so well known to those who have examined stalactitic grottoes with their stalagmitic floors in the clefts and caverns of limestones, or still more those large masses of Tufa (or soft calcareous stone formed by depositions from water) which have proceeded from calcareous wells. Whenever a portion of the Malachite has been broken off, the interior is seen to consist of a number of fine laminæ (a fasciculus of radio-concentric globules), which invariably arrange themselves around the centre on which they are formed, and are adapted to every sinuosity of the pre-existing layer, here presenting a dark line, there a bright and light one, just as the solution of the moment, the day or hour, happened to be more or less impregnated with colouring matter. . . . On the whole we are disposed to view it as having resulted from copper solutions emanating from all the porous loose surrounding mass, and which, trickling through it to the lowest cavity upon the adjacent rock, have in a series of ages produced this wonderful subterranean incrustation."

Enormous solid masses of Malachite are sometimes found, particularly at the mines of Nizhni Tagilsk, Ekaterinburg, on the Siberian side of the Uralian Mountains, the property of Prince Demidoff, one block measuring no less than 18 feet by 9, whilst the complete bed opened up yielded about half a million pounds weight of pure Malachite. It is easily cut, and takes a high .polish, so is largely used as a veneer in the manufacture of works of art, such as tables, vases, mosaics, etc. Entire rooms in several European palaces are veneered with Malachite, while complete sets of furniture have been exhibited by the manufacturers for the Russian Governments. The magnificent Malachite gates and suites of furniture, including chairs, tables, secretaires, etc., exhibited in the Exhibition of 1851, which attracted so much attention, will be remembered by many.

A chamber at Versailles in the Grand Trianon is furnished with mantelpieces, pier and centre tables, basins, ewers, and enormous vases, of Malachite, a present to Napoleon by the Emperor Alexander. Many fluted Corinthian columns made of Malachite are to be seen in the churches in St. Petersburg, and magnificent ornaments are in the palace of the Emperor of Germany at Potsdam.

Her Majesty the Queen possesses an extraordinary vase, placed at one end of the Waterloo chamber, Windsor Castle, about 12 feet high, the gift of the late Czar Nicholas; and Mr. Panmure Gordon, of Loudwater, another not quite so large, as well as an extensive and most beautiful series of other objects vaneered in the same material, which he has assiduously collected for many years both in Siberia and London. If *sparingly* used, Malachite ornaments are of the greatest utility in decoration, lighting up a corner here and there, and always presenting a delicious cooling, even if crude, appearance.

It is used for veneering vases, columns, furniture of every description, including cabinets, chairs and tables, also snuff-boxes, knife-handles, and veneering purposes generally, as well as in jewelry of every description. In mosaic work it is perhaps of the greatest utility, mixed with black and other marbles. Its intrinsic value does not fluctuate much, ranging from ten shillings to a sovereign per pound, according to colour and shade, which is divided into *ordinaire, claire, pale*, and *foncée*, as well as *ronde, longue*, and *tachetée*. The darker colour is the cheapest, whilst the lighter and concentric variety brings the higher price.

MALACHITE. *Molochitis*, pt. Pliny : χρυσόκολλα pt. Theophr. μαλάχη or μολόχη = marsh mallow. Green carbonate of copper. Theophrastus called it pseudo-emerald.

Monoclinic. Seldom found crystallized, but occasionally at the Burra-burra Mines, Australia. Commonly massive, botryoidal, delicately fibrous and banded. Cleavage to the basal plain perfect. Fracture subconchoidal, uneven, brittle.

Hardness 3·5 to 4·0. Specific gravity 3·928 to 4·03, Damour. Colour bright green. Streak paler. Generally opaque.

Composition : Basic Carbonate of Copper.

Carbon Dioxide	...	19·9
Cupric Oxide...	...	71·9
Water... ...	...	8·2
		100·0

Another analysis, by Struve, of a Malachite, from Gumeschewskoi in the Urals, yielded :

Carbon Dioxide ...	...	19·08
Protoxide of Copper	...	72·11
Water	...	8·81
		100·00

Associated often with other ores of copper in large quantities in Siberia, and at the Burra-burra Mine in Australia; also at many localities in America, as well as in the Tyrol, France, Spain, and associated with other ores in Cumberland, Cornwall, and Cork and Limerick in Ireland.

No. 628. MALACHITE. Green carbonate of copper. A large section, cut perpendicularly and exhibiting distinctly in layers the successive deposition in zones of the mineral by cupric percolation. Of a beautiful light and dark green colour in wave-like bands, and at top and bottom exhibiting botryoidal concretions. A good thick section, fastened upon a base of slate. 9 by 6¼ ins.

Copper Mines of Nizhni Tagilsk, Siberia.

Collection of His Imperial Highness the late Duc Nicholas, of Leuchtenberg.

No. 629. MALACHITE. Green carbonate of copper. Banded, of a darkish bright green in beautiful zones; exhibiting Mammillated structure. Beautifully polished. 5½ by 3½ ins.

Nizhni Tagilsk, Siberia.

No. 630. MALACHITE. Green carbonate of copper. A fine massive reniform and botryoidal specimen of a beautiful light green colour. A fine solid piece of exceptionally good quality. Polished partly. 4½ by 3 ins.

Nizhni Tagilsk, Siberia.

Collection of H.I.H. the late Duc Nicholas of Leuchtenberg.

No. 631. MALACHITE. Green carbonate of copper. Dark and light botryoidal specimen. Very beautiful. Partly polished. 3¾ by 3 ins.

Nizhni Tagilsk, Siberia.

No. 632. MALACHITE. Green carbonate of copper. In concentric and botryoidal circles and zones, associated with deep coloured Azurite (blue carbonate of copper), tipping the Malachite, with Chrysocolla of a lighter green—hydrous silicate of copper. Beautifully polished on one side. 3¼ by 3 ins.

Morenci, Arizona.

A beautiful combination of three varieties of copper.

No. 633. MALACHITE, associated with Azurite, blue carbonate of copper, of a beautiful dark blue with a little pale green Chrysocolla, the silicate of copper. 4 by 3¼ ins.

Morenci, Arizona.

No. 634. MALACHITE. Cross section of a Stalactite in botryoidal concretions embedded in beautiful blue Azurite, the blue carbonate of copper. Polished on one side. 1¾ by 1½ ins.

Morenci, Arizona.

No. 635. MALACHITE associated with Chrysocolla. Pale in botryoidal concretions with Azurite. Polished on one side. 5½ by 3¾ ins.

Morenci, Arizona.

SILICATES.

A. ANHYDROUS.

Feldspar Group.

FELDSPAR is a most important mineral, consisting essentially of silicates of alumina, potassium, soda, magnesia and lime, generally crystallised or occurring as a constituent of the crystalline rocks. Its ordinary variety, Orthoclase, occurs in granite, gneiss, syenite, as well as porphyry, trachyte phonolite, and other rocks. Another variety, Labradorite, is an essential constituent of the basic rocks, generally associated with minerals of the Pyroxene or Amphibole Groups. As stated in the Chapter on Agate, page 68, it is an ingredient of the Melaphyre of Oberstein, which would easily become decomposed when acted upon by carbonated waters. Dana points out that "when the infiltrating waters contain traces of carbon dioxide, the Feldspar acted on first loses its lime, if a lime feldspar like Labradorite, by a combination of the lime with this acid ; next its alkalies are carried off as carbonates, if the supply of carbonic acid continues, or otherwise, as silicates in solution. The change thus going on ends in forming kaolin or some other aluminous silicate. The carbonate of soda, or potash, or the silicate of these bases, set free, may go to the formation of other minerals—the production of pseudomorphic or metamorphic changes—and the supplying fresh and marine waters with their saline ingredients. When the change is not carried on to the exclusion of the ferrous bases, certain zeolites result, especially as Bischof states, when Labradorite (the lime feldspar) is the Feldspar undergoing alteration. When the waters contain traces of a magnesian salt— a bicarbonate or silicate—the magnesia may replace the lime or soda, and so lead to a steatitic change, or to a talc when the alumina is excluded ; and when augite or hornblende, it may give origin to steatite." The author has quoted the foregoing *in extenso* to show how easily Feldspar can be altered through infiltrating waters containing carbon dioxide in solution, depositing finally free silica which may form the amygdaloidal Agates at Oberstein. The Feldspars crystallize in two distinct systems, the Monoclinic and Triclinic. The specimens represented in the collection consist of a variety of Orthoclase—Perthite—crystallizing in the Monoclinic system, and the beautiful pale blue metallic Microclines and lovely chatoyant Labradorites, which belong to the Triclinic.

FELDSPAR, Felspath *Germ.* Orthoclase, and varieties Monoclinic. Microcline, and Labradorite Triclinic. Fracture conchoidal to uneven. Hardness 6·0. Specific gravity 2·39 to 2·62. Brittle. Composition : A silicate of aluminium and potassium, with soda, magnesia, and lime.

Orthoclase*
(from the Greek ὀρθός = straight, and κλάσις = cleavage).

(a) *Perthite.*

No. 636. PERTHITE, so named from its locality. A flesh red-coloured aventurine Feldspar, consisting of Orthoclase, presenting a metallic chatoyancy. Interlaminated with Albite. Well polished. 3 by 2½ ins.

An analysis made of Perthite from Bathurst, Perth, by Hunt, yielded :

Silica ...	66·44
Alumina ...	18·35
Ferric Oxide	1·00
Potassium ...	6·37
Soda... ...	5·56
Lime... ...	0·67
Magnesia ...	0·24
Loss by Ignition	0·40
	99·03

Bathurst, near Perth, Quebec, Canada.

* *Thomson,* "*Phil. Mag.*," 22, 189 ; 1843.

No. 637. PERTHITE. A flesh-red aventurine Feldspar consisting of Orthoclase inter-laminated with Albite, presenting a metallic chatoyancy.　2¾ by 2 ins.

Bathurst, Perth, Quebec, Canada.

No. 1. Microcline.

MICROCLINE, Mikroklin, *Breith.* An Orthoclase Feldspar, from μικρός= little, and κλίνειν to slant, occurring in cleavable masses in the Zircon Syenite of Fredriksvärn, Laurvik, and Brevig, Norway. Massive. Brittle. Hardness 6·65. Specific gravity 2·54 to 2·57. Pale cream-yellow. Triclinic. Fracture uneven. Brittle. Exhibits a beautiful play of colour of a pale moonlight metallic blue at a certain angle.

Formula $K\ AlSi_3O_8$ or $K_2O.\ Al_2O_3.\ 6SiO_2$ to

Silica	...	64·70
Alumina	...	18·40
Potash	...	16·90
		100·00

With a trace of Sodium according to Dana.

Another analysis of this chatoyant Microcline, from Fredriksvärn, gave Gmelin :

Silica	...	65·18
Alumina	...	19·99
Ferric Oxide		0·63
Potassium	...	7·03
Soda...	...	7·08
Lime...	...	0·48
Loss by Ignition	...	0·38
		100·77

No. 638. MICROCLINE. Exhibits beautiful blue metallic moonlight tints at two or three distinct angles. Much cracked all over the surface. Polished.　5 by 4¼ ins.

Zircon Syenite of Fredriksvärn, Norway.

No. 639. MICROCLINE. Exhibits beautiful blue metallic moonlight reflections at two distinct angles. Well polished. 4¾ by 4½.

Zircon Syenite of Fredriksvärn, Norway.

No. 640. MICROCLINE. A most beautiful solid specimen exhibiting a pale (nearly white) metallic moonstone tint. Presenting a lovely white velvet appearance. Well polished. 3¾ by 2¾ ins.

Zircon Syenite of Fredrisksvärn, Norway.

No. 641. MICROCLINE. Exhibiting distortion at the junction of three distinct angles, which reflect beautiful blue metallic moonlight tints in three different ways. Well polished. 4½ by 3 ins

Zircon Syenite of Fredriksvärn, Norway.

No. 2. Amazonite.

No. 642. AMAZONITE or AMAZON STONE. So named from the first locality in which it was found, the River Amazon. Crystallized. A variety of Microcline in a group of bright verdigris-green crystals used for inlaying, and by the Assyrians for cylinders. 2½ by 2 ins.

Granite of Pike's Peak, Colorado.

An analysis of a Siberian Amazonite yielded Abich:

Silica..	65·32
Alumina	17·89
Ferric Oxide and Oxide of Copper	0·30
Potassium	13·05
Soda	2·81
Lime...	0·10
Magnesia	0·09
Manganese Sesquioxide ...	0·19
	99·75

This mineral would be of great value were the colour homogeneous through the crystals—in the centre they are unfortunately white.

LABRADORITE.

OF all the minerals classified under the group Feldspar, none are so beautiful as this variety with its lovely chatoyant reflections which occur in every prismatic colour. It is said to have been first brought from the Isle of St. Paul, on the west coast of Labrador, by Mr. Wolfe, a Moravian missionary, about the year 1770, and was known under the name of *Labradorstein*, as well as chatoyant, opaline, and Labrador-feldspar, and subsequently by Bishop Launitz in 1775, who introduced it into Europe. It is an essential constituent of many rocks, in which it is associated with Hornblende, Augite, Diallage, and Hypersthene.

Soon after its discovery a Dr. Anderson having minutely described its lovely changeable chatoyant colours, it became at the time of great value, twenty pounds and more being paid for small specimens.

At the present day it is not so valuable in the rough, but when cut at certain angles, so as to catch the play of light in a required direction, which is a matter of great difficulty, it is of great value. The author is informed by the captain of a vessel who annually brings blocks from Labrador, that the finest specimens are not found on the coast, but about one hundred miles inland, and that nearly all he procures has to be conveyed to the boat over a very rough country by sleighs drawn by dogs, adding greatly to its commercial value.

Lapidaries as a rule examine pieces of rough stone, note the angle on which the reflection is exhibited, and then cut slices off accordingly, having a great preference for the blue-coloured specimens

This, the easiest mode of procuring examples of resplendent Labradorite has the disadvantage that such specimens will only show the colour by being held and turned so as to catch a particular angle, and if placed horizontally as a rule lose all colour.

An eminent authority stated that to cut ten or a dozen specimens, place them *horizontally* so as to procure a *continuous* colour would be next to impossible. The author, however, after many experiments and failures, succeeded in overcoming the difficulty, and produced a rectangular table-top measuring thirty-three inches by twenty-five, the centre of which was composed of magnificent celestial-blue-coloured specimens of Labradorite, encircled by a *filet* about half an inch diameter of flaming red, surrounded by a border of specimens about three inches square of different but most exquisite hues, the whole top in *one direction* looking like an ordinary grey marble table-top, but in *another*, although composed of hundreds of different specimens presenting one *continuous* blaze of light, and exhibiting, perhaps, the most brilliant combination of chatoyant reflections ever massed together in so small a space. This remarkable *chef d'œuvre* of the author is now the property of His Grace the Duke of Westminster, K.G., and is in the library of Eaton Hall. The following extract from the *Pall Mall Gazette*, describing the table when on view, will be of interest :

" What a pity that Mr. Ruskin is not well enough to come up to town to see at Mr. Bryce-Wright's the wonderful table-top of lime-feldspar from the coast of North America, as its owner styles it, but which adequately to describe would tax all the wealth of the ' Master's ' luxuriant vocabulary. As the beholder first sees it, this extraordinary table presents the appearance of a sombre grey marble ; but let the eye catch it at a certain angle, and it gleams forth in blues and reds, for a comparison with which one must literally resort to the metallic lustre of the feathers of the humming bird and the brilliancy of the most precious gems. The difficulty which

O

the skilful artificer had to overcome consisted in the necessity of cutting each of the numerous separate and selected stones which make up the mosaic of the table in such a way that the azure or crimson gleam from each of them should converge in one blaze of brilliancy upon the eye of the spectator.

*   *   *   *   *

" The quantity of material which had to be examined and tested in order to get a sufficiency of first-rate specimens for the composition of this small slab was more than a ton's weight. A further trouble has been that the nature of the substance prevents it being cut with a wheel by machinery and it has been requisite in every case for the lapidary to use a fine steel wire strung upon an archaic and clumsy-looking bow, with which, when he has painfully discovered the necessary angle at which to work, he has with the aid of emery laboriously to sever the material. *Surely so much splendour was never before contained in a surface of considerably less than a square yard.* Passers-by should see the sight for themselves, especially if they be millionaires and do not mind going as far as four figures in the purchase of a table-top."

A wonderful and unique casket, the top, front, and sides of which are of Labradorite, was also made by the author for the distinguished collector Mr. Alfred Morrison, in such a manner that, placing the casket so that the light falls upon its front, the top exhibits, with the front and sides, remarkable brilliant chatoyant reflections in one blaze of light ; the specimens being cut at different angles according to the position assigned to them, and placed so as to ensure *continuous uninterrupted* colour, both upon the front, top, and sides.

The specimen No. 646 is one of the most remarkable known, as it exhibits the *crystalline form* of the mineral, in exquisite chatoyant colours of old gold and an indescribable blue which are not often met with. Nos. 647 and 649 also are worthy of particular attention.

LABRADORITE. Labrador Feldspar. *Labradorstein.*

Triclinic. Colour grey, brown, and greenish, also white (but rare) and glassy. The cleavable varieties exhibit a remarkable chatoyant play of prismatic colours. The blue colour Vogelsang regards as a polarization phenomenon due to its lamellar-structure, whilst the golden or reddish is due "to the presence of black acicular microlites and yellowish-red microscopic lamellæ, or to the combined effect of these with the blue reflections.* Translucent to subtranslucent. Lustre pearly, passing into vitreous.

Composition from an analysis by Thomson :

Dioxide of Silicon	...	...	...	47·57
Alumina	...	...	...	29·65
Lime ...	...	...	...	9·06
Magnesia	...	...	...	0·40
Soda ...	...	...	...	7·60
Water ...	...	...	...	1·28
Ferrous Oxide...	...	...	...	3·57
				99·83

Another analysis by Klaproth of a specimen from the Isle of St. Paul, coast of Labrador, yielded :

Dioxide of Silicon	55·75
Alumina	26·50
Ferric Oxide ...	1·25
Lime	11·00
Soda	4·00
Loss by Ignition...	0·50
	99·00

Hardness 6. Specific gravity 2·68 to 2·76. An essential constituent of many rocks.

The magnificent reflections are not due to any chemical constituent, but to a peculiarity in its intimate or minute lamellar structure, as mentioned by Vogelsang which, when viewed under the microscope, reveals the existence of a cleavage formation resembling a complex system of delicate gratings, transverse to the brachydiagonal section.

* *"A System of Mineralogy," Dana,* 1892, *p.* 334.

No. 3. *Labradorite.*

No. 643. LABRADORITE or LABRADOR FELDSPAR, so
called from the first locality in which it was found, with very
fine striations on the cleavage surface. Exhibiting brilliant
combinations of celestial dark blue chatoyant reflections,
associated with mottled green, yellow, purple, and violet.
Cracked on the surface, which is traversed with grey white
lines. Highly polished. $7\frac{1}{2}$ by $5\frac{1}{4}$ ins.

> Occurs in the Archæan Rocks of the Island of Paul,
> Labrador.

No. 644. LABRADORITE or LABRADOR FELDSPAR. A
magnificent thick specimen of celestial-blue chatoyant reflections with rainbow coloured bands of old gold, violet,
bluish-green, and yellowish-green, with white Feldspar, in
which there is scattered blue, orange, and violet tints, with
Magnetite exhibited on both sides of the stone. $6\frac{1}{2}$ by 5 ins.

> Archæan Rocks of the Coast of Labrador.

One of the finest blocks of Labradorite known.

No. 645. LABRADORITE, showing contortion of the rock with
chatoyant reflections at different angles. The centre is of a
celestial-blue tint, associated with yellow and violet. When
viewed at other angles it exhibits strata of blue, and blue
with red. Cracked in one direction. Well polished and
mounted on slate. $6\frac{3}{8}$ by $4\frac{3}{8}$ ins.

> Coast of Labrador.

No. 646. LABRADORITE or LABRADOR FELDSPAR. Exquisite specimen, exhibiting trimetric form of the mineral, the
colours arranging themselves at one angle of light into a
centre of old gold, with a violet coloured band, surrounded
by irregular celestial-blue, with greenish-yellow and blue, all
in exquisite chatoyant reflections, whilst at another angle the
centre is of a beautiful sage and pale green crossed by black
parallel lines also following the crystalline trimetric form.
$5\frac{4}{4}$ by 5 ins.

> Archæan Rocks of Labrador.

An extraordinary specimen, exhibiting as it does its crystalline form
distinctly upon the surface. The author believes this specimen to
be quite unique.

No. 647. LABRADORITE or LABRADOR FELDSPAR. A thick specimen exhibiting beautiful chatoyant reflections at different angles. A wide band of old gold with a reddish tinge tipped with celestial-blue on one angle, with deep blue associated with green and red, mottled at another angle. Well polished. 6 by 4 ins.

Coast of Labrador.

An extremely brilliant and unusually fine specimen, the reflections being broad and most decided.

No. 648. LABRADORITE, in beautiful dark blue with yellow, red, violet, and green chatoyant reflections, somewhat mottled, well spread all over the stone. Grey and much cracked in one direction. Thin specimen. Well polished. 5½ by 4¾ ins.

Isle of St. Paul, Labrador.

No. 649. LABRADORITE, exhibiting a beautiful play of colours at two different angles, one of vivid celestial-blue with yellow, violet, and red, another of deep vivid celestial-blue with yellow, yellowish-green, with a rare violet "peacock-copper" iridescent reflection in some parts. A contorted specimen. Attached to slate. Well polished. 5 by 5½ ins.

Isle of St. Paul, Labrador.

A most brilliant example.

No. 650. LABRADORITE, of a most lovely and intense dark blue tipped with vivid green and yellow at one angle, exhibiting at another dark, nearly black, reflections, with bands or strata of blue distinctly separated. 5 by 5½ ins. and ¾ in. thick.

Isle of St. Paul, Labrador.

No. 651. LABRADORITE. LABRADOR FELDSPAR. Magnificent continuous vivid purple-violet reflection exhibited over the whole stone at one angle, grey, much cracked, with white lines at another mounted on slate. 4½ by 4¼ ins.

Isle of St. Paul, Labrador.

A very rare colour to find wholly disseminated over a large surface.

No. 652. LABRADORITE. A magnificent specimen exhibiting a beautiful peacock-blue with violet, orange, and chatoyant reflections at one angle; at another exhibiting beautiful blue, green, and violet tints at one side in layers. General appearance grey, much scratched and with dappled white markings. Polished on both sides. 7¼ by 4¼ ins. by ¾in. thick.

Isle of St. Paul, West Coast of Labrador.

No. 653. LABRADORITE, exhibiting a lovely peacock-blue colour, arranged in parallel layers, with yellow and red chatoyant reflections. It has also over it a peacock-copper iridescence. Well polished. 4 by 3½ ins.

Isle of St. Paul, West Coast of Labrador.

No. 654. LABRADORITE, of a whitish grey, but exhibiting beautiful blue celestial chatoyant reflections in patches with a little yellow and green. Magnetite can also be detected in spots, grey in one direction and much cracked. Well polished. 6¼ by 5¾ ins.

Isle of St. Paul, Labrador.

No. 655. LABRADORITE, celestial-blue chatoyant reflections dappled with yellow, purple, red, and green. Exhibits also the rare peacock copper iridescence. Attached to slate. Well polished. 6 by 3½ ins.

Isle of St. Paul, Labrador.

No. 656. LABRADORITE, a large slab in light and dark celestial-blue reflections. Exhibits white streaks and fine striations. A thin specimen. Well polished. 7½ by 5¼ ins.

Isle of St. Paul, West Coast of Labrador.

No. 657. LABRADORITE, in most beautiful violet and yellow reflections with blue. Exhibits an unusual white metallic reflection at one angle. Well polished. 4¼ by 3¼ ins.

Isle of St. Paul, Labrador.

No. 658. LABRADORITE, exhibiting a most lovely golden-green reflection, tipped with map-like celestial-blue. The golden-green is in continuous colour all over the surface, which at the top is banded, and much cracked in appearance. Well polished. 4 by 2¾ ins.

Coast of Labrador.

No. 659. LABRADORITE, with a lovely play of chatoyant colours in irregular patches of vivid red, purple, old gold and blue reflected at one angle. Exhibits white metallic reflections at another. Well polished. 5½ by 2⅜ ins.

Labrador.

No. 660. LABRADORITE, reflecting a lovely chatoyant, light peacock-blue colour well disseminated, with a little vivid violet and yellow in map-like form. Grey, with white metallic reflections at another angle. Well polished. 5 by 3 ins.

Isle of St. Paul, West Coast of Labrador.

No. 661. LABRADORITE, with a beautiful vivid celestial-blue reflection at one angle, nearly continuous over the stone. It has, also, a white dappled appearance at another angle. Well polished. 3¼ by 2½ ins.

Isle of St. Paul, Labrador.

No. 662. LABRADORITE, reflecting deep vivid red orange-yellow and violet chatoyant tints in one direction, in another white metallic reflections. Well polished. 3¼ by 2⅜ ins.

Labrador.

No. 663. LABRADORITE, with a beautiful play of pea-cock-blue colour all over the stone ; grey and cracked at one angle of light. Polished. 3¼ by 2¼ ins.

Coast of Labrador.

No. 664. LABRADORITE, slab of, with a beautiful chatoyant play of celestial-blue at one angle, grey with white patches at another. Polished. 3 by 3 ins.

Coast of Labrador.

No. 665. LABRADORITE, slab, with beautiful reflections in green and gold, not very distinct. Polished. $3\frac{1}{4}$ by $2\frac{3}{8}$ ins.

Coast of Labrador.

No. 666. LABRADORITE, reflecting celestial-blue, with vivid yellow and violet colours. Grey and dappled white in one direction. Polished. $2\frac{3}{4}$ by $2\frac{1}{4}$ ins.

Coast of Labrador.

No. 667. LABRADORITE, slab, exhibiting bands of curved blue in one direction, and continuous peacock-blue in another ; also reflections of a metallic white. Well polished. $3\frac{3}{8}$ by $1\frac{1}{2}$ ins.

Coast of Labrador.

No. 668. LABRADORITE, exhibiting chatoyant colours of a sage green and peacock-blue in strata or bands. Polished. $3\frac{1}{2}$ by $1\frac{1}{2}$ ins.

Isle of St. Paul, West Coast of Labrador.

No. 669. LABRADORITE, with reddish golden reflections associated with violet ; white metallic reflection at another angle. Polished. $2\frac{1}{4}$ by $2\frac{1}{8}$ ins.

Isle of St. Paul, Labrador.

No. 670. LABRADORITE. Rectangular specimen. Exhibiting at one angle irregular patches of green and blue in layers associated with bronze-coloured Hypersthene. A dark looking specimen mounted on a slab of iron. $4\frac{1}{4}$ by $3\frac{1}{4}$ ins.

Farsund, Norway.

PYROXENE GROUP.

No. 1. Enstatite, Bronzite.

No. 671. ENSTATITE or BRONZITE (from the Greek ἐνστάτης = an opponent). Rectangular specimen. Diallage metalloidal. Olive green base with light bronze and pearly reflections. Well polished. 4½ by 3¼ ins.

An analysis by Damour yielded :

Silica...	56·70
Magnesia	33·61
Ferrous Oxide	7·72
Alumina	0·60
Water	1·04
	99·67

Vosges, France.

No. 2. Hypersthene, Labrador.

No. 672. HYPERSTHENE (from the Greek ὑπέρ = very, and σθένος strength, meaning *very tough*). A magnificent slab of a chatoyant metallic pinchbeck-brown or bronze colour, continuous over the whole stone. Cracked over the surface, but not, as usual, of a white or different colour to the principal body of the stone, but of the same bronze colour. Specific gravity, 3·402. Well polished. 9 by 4¼ ins.

An analysis yielded Remelé [*] :

Silicon Dioxide ...	49·85
Alumina	6·47
Ferric Oxide	2·25
Ferrous Oxide	14·11
Manganese ...	0·67
Magnesia ...	24·27
Lime... ...	2·37
	99·99

Isle of St. Paul, West Coast of Labrador.

[*] " *Ber. Ch. Ges.*," 1, *p.* 145, 1868.

No. 673. HYPERSTHENE. Rectangular with BRONZITE, a variety of Enstatite (Diallage metalloidal), associated with a rare coloured blue Labradorite in irregular strata, tinged slightly here and there with red and violet. Well polished all over. 6 by $4\frac{1}{8}$ ins.

Norway.

A most extraordinary specimen and extremely difficult to describe, the ground being of a metallic black, in which the Hypersthene is embedded, whilst the Bronzite almost imperceptibly at the line of junction runs into it—the Hypersthene.

No. 3. *Jadeite of Damour.*

JADEITE of Damour.* Nephrite or Jade, *Part.* A soda spodumene. An apple-green variety of Jade classified with the Pyroxenes. It is used extensively in China for ornaments under the name *feitsui.* Massive. Fracture splintery. Extremely tough. Formula $NaAl\,(SiO_3)_2$, or $Na_2O\,Al_2O_3\,4SiO_2$, with composition :

Silica ...	... 59'4
Alumina	... 25'2
Soda ...	... 15'4
	100'00

Essentially a meta-silicate of sodium and aluminium corresponding to spodumene.

Analysis by Damour yielded :

Oxide of Silicon	57'99
Alumina ...	20'61
Ferric Oxide	2'84
Lime... ...	4'89
Magnesia ...	3'33
Soda... ...	9'42
Potassium ...	1'50
	100'58

Hardness 6'5 to 7'0. Extremely tough. Specific gravity is very high, like that of Zoitite, 3'16 Damour, which is higher than the true Nephrite, 2'95 to 3'00.

* *"Comptes Rendus,"* 56, 861, 1863.

No. 674. JADEITE or NEPHRITE, *Part. Damour.** Chloro-melanite. An apple-green colour of two shades, one bright green and the other of a greenish-blue. Evidently cut from a water-worn boulder, and exhibiting exterior. Well polished. 3 by 2½ ins.

Durance, France.

No. 4. Rhodonite.

No. 675. RHODONITE, ρόδον =a rose, a bisilicate of manganese. Hardness 5·50 to 6·50. Specific gravity 3·40 to 3·68. A manganese meta-silicate, with the formula: $MnSiO_3$, or $MnO \ SiO_2$, equal to—

Silica	...	45·9
Manganese protoxide	...	54·1

100·0

Iron, calcium, and occasionally zinc, replace part of the manganese.

Another analysis by Berzelius yielded :

Silica...	...	48·00
Manganese Dioxide	...	49·04
Lime... ...	...	3·12
Magnesia ...	...	0·22

100·38

A very large rose-red specimen, distinct in colour, with a few black patches disseminated over it. Well polished. 12 by 7 ins.

Ekaterinburg, Siberia.

No. 676. BALL cut out of rose-pink Rhodonite, with a black marking caused by oxide of manganese near the hole. Diameter 1½ ins. Cut and polished at Oberstein.

Ekaterinburg, Siberia.

No. 677. BALL cut out of rose-pink Rhodonite. Diameter 1½ ins. Cut and polished at Oberstein.

Ekaterinburg, Siberia.

* *"Comptes Rendus,"* 56, 86, 1863.

AMPHIBOLE GROUP.

Nephrite, Jade or Greenstone.

NEPHRITE, or, as it is commonly called, JADE, is derived from the Greek word νεφρός, meaning a kidney, in allusion to its having been considered a cure for diseases of that organ by the Maories or New Zealanders, who suffered terribly, from that disease when Europeans first visited their island. It is called *Yu* or *Yu-shih* (Yu stone) by the Chinese, by whom it is greatly valued, and who work it most elaborately into vases and ornaments of every description. The Maories in their language call it *Pounamu* or *Poenamu*. It is also known as "Axe stone" or "Hatchet" stone, and in old books as green talc, by which name Captain Cook heard marvellous stories of it in his winter station at Queen Charlotte's Sound. It is a mineral which has particularly arrested the attention of the mineralogical world, through the discovery of implements made of it (and its varieties) in the lacustrine habitations or prehistoric lake-dwellings of Switzerland, in various parts of France, as well as among the ancient monuments of Central and South America; whilst it was believed, that the only localities known to exist were Burmah, China, Siberia, with New Zealand, New Caledonia, and other islands in the Pacific.*

Hence arose a great question, From whence did prehistoric man in Switzerland, France, and America procure this material? Various hypotheses were immediately advanced to show that Jade must have been procured by contact of early

---

* See a paper on "The Source of the Jade used for ancient implements in Europe and America," by Prof. F. W. Rudler, Anthropological Institute, 1891.

races, with the New Zealanders and inhabitants of the Pacific, as well as Central Asia. Professor Fischer of Freiburg in Baden, a noted mineralogist, believed entirely in its "exotic" existence, and left no stone unturned to prove his theory, publishing a most exhaustive treatise on it in 1880.* Since that date, however, it is found that the geographical distribution of Jade is not so limited as supposed, and it has been found at several European localities. Monsieur Damour, the great French savant and chemist, first pointed out that the mineral called Jade really consisted of two varieties, Jade and another, to which latter he gave the name of Jadeite.† Shortly afterwards he found a pebble or boulder of the Jadeite variety on the Lake of Geneva, and later a larger boulder was found near Leipzig, which Professor Fischer accounted for by suggesting they were accidental fragments dropped probably during the migration of some prehistoric tribe. But further discoveries took place ; Jade pebbles were said to be found in the drift near Potsdam ; then three specimens on three different occasions were found in Styria, followed by two varieties, *in situ*, in Silesia, as well as at Durance in France (see No. 655). Thence followed the discovery of Jade in British Columbia,‡ and also of large quantities of it, *in situ*, by Lieutenant G. M. Stoney at the Jade Mountains north of the Kowak river, Alaska ; so that the balance of evidence is, as Professor F. W. Rudler remarks, "that Jade is for the most part indigenous to the countries in which the implements occur."

It is, however, principally with the true Nephrite or "Greenstone" we have to deal, which is so well represented in the collection. It is found of two tints of green, light and dark,

---

* "*Nephrit und Jadeit, nach Ihren Mineralogischen Eigenschaften sowie nach Ihrer urgeschichtlichen und ethnographischen Bedeutung.*" *Von Heinrich Fischer, Stuttgart,* 1875 *and* 1880.

† "*Comptes Rendus,*" 56, 861, 1863.

‡ "*Notes on the occurrence of Jade in British Columbia, and its Employment by the Natives.*" *By George M. Dawson, D.Sc., F.G.S., etc., Canadian Record of Science, Vol.* ii., *No.* 6, *April,* 1887.

as well as mottled, on the west coast of South Island, New Zealand, generally in boulders from the gravel deposits of two large valleys. The dark tint is called by some of the Maories, *Pounamu rau-karaka*, after the colour of a variety of a dark green laurel-leaf, called by them Rau Karaka, from Rau—a leaf and Karaka=a laurel—which is known as the *Coryno-carpus lævigata;* whilst the light green is called *Pounamu puka-puka*, after the *Brachyglottis repanda*, neither of which is, however, a true laurel.* A very rare variety is called *Tang-i-wai* (tear-water), possessing an opalescent chatoyant sheen, which is in reality a precious Serpentine, not a Jade. Nephrite is known throughout New Zealand as " Greenstone," the common Maori name proper being *Pounamu.*

" Pounamu † was one of the sons of the great Polynesian deity Tangaroa (Lord of the Ocean), who was the son of Rangi (Heaven) and Papa (Earth). Tangaroa married Te Anumatoa (the Chilly god), who became the mother of four gods, all of the fish class, of whom Pounamu was one. The substance *Pounamu*, it is stated, was formerly generated inside a fish (the shark), and only became hard on exposure to the air. The stone *Pounamu* was classified as a fish by the Maories." The following is one of many mythical Maori histories attached to Jade, told by Mr. Chapman :

" *Tamatea-pokai-whenua*, a celebrated ancestor of Maori tribes, in addition to his faithful wives, had three : *Hiner-aukawa, Hinerauharaki,* and *Te Kohiwai,* who deserted him He sailed right round South Island in search of them, naming the rivers and headlands as he passed.

---

* *Some doubt having been expressed by a high authority as to this Maori nomenclature, the author addressed a letter to the Director of Kew Gardens, and received an answer as follows : " Royal Gardens, Kew, 26th Sept., 1893. Sir,—I can only say that you are probably right.* Corynocarpus *certainly bears the name Karaka, and puka-puka is one of the native names of* Brachyglottis repanda. *Neither is a true laurel. Yours faithfully, W. T.* THISELTON-DYER."

† *See Paper by F. R. Chapman, read before the Otago Institute, Oct. 14th, 1891. " Trans. and Proc. N.Z. Inst.," 1891, Vol. xxiv.*

" Though he listened for every sound indicative of their presence, it was not until passing up the west coast he reached the Arahura river that he heard their voices. He failed, however, to discover his wives, for he did not know their canoe had been upset here, and they and all the crew had been transformed into stones. His slave happening to burn his fingers while cooking some birds they had killed, impiously licked them, urged by pain. He was instantly turned into the mountain *Tumuaki*, which stands there still, and as a consequence *Tamatea* never found his wives. Since then, the flaws which sometimes discolour the best kinds of Greenstone are called *tutekoka*, the excrement of the birds the slave was cooking when he did this wrong." There is also the legend of " *Poulini* and *Whaiapu* " mentioned in Sir George Grey's " Polynesian Mythology,"as well as many others.

It is generally looked upon as an exceedingly hard stone, but this is not a fact. Its hardness is only 6·0 to 6·5, under that of Quartz or Silica. It possesses, however, the degree of frangibility termed toughness to a remarkable extent. The Maories made Jade into implements and gods, the principal implement being called the " *Mere* " or " *Patoo-patoo*," which is a long oval weapon with a short handle, perforated in the centre, cut from one piece. They also cut charms or gods out of it, called " *Hei-Tikis* " or " *Tikis* "; they are broad and flat, semi-polished figures, squatting upon their haunches with their arms crossed over the breast, the eyes large and round, and are generally inlaid with shell cut from the interior of the Haliotis; the mouth is very wide and open. They also make eardrops, long and short, *Kurukuru Rapen Mako*, some like sharks' teeth, termed " *Mako*," as well as axes (*tokis*), and other weapons.

The author had the pleasure of speaking to King Te Whaiao during his residence in England, who explained that the " *Meres*," as well as " *Hei Tikis* " were formerly title deeds to tracts of land.

The sacking of the Summer Palace in Pekin caused many magnificent specimens to be brought to Europe of the Chinese Jade, most elaborately carved, a wonderful collection being made by the late Arthur Wells, Esq., of Nottingham, a magnificent series of which he presented to the South Kensington Museum, where they are now exhibited. The rarest kind of Jade is of a grass-green, and translucent. A small cup, not six inches in diameter, in the collection of Mr. Alfred Morrison, is valued at 500 guineas, in such high estimation is this variety held. A collection of Jade ornaments from China and India, which bids fair to rival all others known, has been made by Mr. Heber R. Bishop of New York, who is publishing an exhaustive and, to all accounts, wonderful work upon Jade and Jade ornaments.

The large section No. 684 in the Derby Collection is one of the largest and thinnest known.

NEPHRITE or JADE. Axe stone. The "Greenstone" of New Zealand. Pietra di hijada. *Pounamu* of the Maories. Old books refer to it as the "Green talc" of New Zealand, by which name Captain Cook heard of it. Lapis Nephriticus, or Kidney stone, from νεφρός = a kidney. *Yu* or *Yu-shih* of the Chinese ; *Pietra d'Egitto* of the antiquaries ; *Nierenstein* of the Germans.

Light to dark green ; sometimes mottled. Occurs massive, either coarse or fine. Breaks with a splintery fracture. Subconchoidal, uneven.

Hardness 6·0 to 6·5. Specific gravity 2·96 to 3·01, according to Dana.

Composition : Silicate of magnesia, lime, and iron. Analysis of a New Zealand specimen :

Silica ...	57·75
Magnesia ...	19·86
Lime	14·89
Ferrous Oxide	4·79
	97·29

With traces of alumina and manganese.

Analysis by M. Damour of Paris :

Silica	...	...	51·70
Magnesia	...	...	23·50
Lime...	...	...	13·09
Ferrous Oxide	...		7·62
			95·91

With 0·95 of alumina and trace of manganese. Loss by ignition 2·42.

No. 1. Nephrite.

No. 678. NEPHRITE or JADE. Axe stone. " Greenstone" of New Zealand. *Pounamu* or *Poenamu* of the Maories. A large dark green slab, mottled white over the whole surface. Rough exterior. One side most beautifully polished. 12⅝ by 7⅜ ins.

North Island, New Zealand.

Cut from a specimen procured by the author from the Maori King Te Whaiao during his residence in England. A most characteristic and extremely fine example.

No. 679. NEPHRITE or JADE. Axe stone. " Greenstone " of New Zealand. *Pounamu* of the Maories. Of a light green colour like the variety of laurel *Brachyglottis repande.* Polished on one side. 12⅝ by 7½ ins.

New Zealand.

A very fine characteristic slab of unusual size exhibiting rough exterior.

No. 680. NEPHRITE or JADE. Axe stone. " Greenstone " of New Zealand. *Pounamu* of the Maories. This is the variety termed *Pounamu puka-puka* after the dark green variety of laurel leaf, called by them *Rau Karaka*—from *Rau*, a leaf, and *Karaka*, a laurel—the *Corynocarpus lævigata.* Of a fine pure dark green. Well polished on both sides. 7¼ by 4¾ ins.

North Island, New Zealand.

A good characteristic dark, pure-coloured specimen of Jade, very slightly mottled.

P

No. 681. NEPHRITE or JADE. Axe stone. "Greenstone" of New Zealand. *Pounamu puka-puka* of the Maories. A fine dark green colour, mottled with white. 6½ by 5½ ins. Most beautifully polished.

North Island, New Zealand.

No. 682. NEPHRITE or JADE. Axe stone. "Greenstone" of New Zealand. *Pounamu puka-puka* of the Maories. A very fine dark green colour, translucent and very sharp at the edges. Cut wedge-shaped with a fine, razor like cutting edge, and well polished on both sides. 4¼ by 4⅛ ins.

New Zealand.

An exceptional beautifully-coloured specimen.

No. 683. NEPHRITE or JADE. Axe stone. "Greenstone" of New Zealand. *Pounamu.* A light-green coloured specimen, somewhat clouded, called after the light-green leaf of a variety of laurel, *Brachyglottis repandc.* Irregular jagged edge. Well polished. 5 by 4 ins.

North Island, New Zealand.

No. 684. ·NEPHRITE or JADE. Axe stone. Greenstone of New Zealand. *Pounamu* puka-puka of the Maories, so called from the dark-green leaf of a variety of laurel called by them *Rau Karaka,* from Rau, a leaf, and Karaka, a variety of laurel (*Corynocarpis lœvigata*).

An extremely large and important section, cut from the centre of an irregular large boulder. Greatly water-worn, as indicated by the outside edge, which is complete and exhibits the entire shape of the stone. Dark apple-green colour, with a little light in irregular cloud-like black markings disseminated through the stone, but not sufficient to detract from its beautiful colour. Stained red at one side, probably with ferrous oxide. Extremely thin and translucent section, very beautiful by transmitted light. Polished on both sides. 15¾ by 12½ ins. but only the eighth of an inch in thickness.

North Island, New Zealand.

This remarkable and abnormal sized specimen is one of the finest and thinnest sections known, and illustrates well the perfection to which the French have advanced in the lapidary art. It required the greatest care to cut such a thin specimen from the centre of a boulder and then to polish it on both sides. It was cut with the aid of the diamond (Boart) with immense hydraulic power, on the Marne, polished with emery, and finally by hand with putty powder.

No. 685. NEPHRITE or JADE, "Greenstone" of New Zealand. Cut into the shape of an open walnut. A good dark green colour. *Pounamu puka-puka* 2 by 2½ ins.

New Zealand.

No. 2. Asbestos.

No. 686. ASBESTOS. ἄσβεστος = inconsumable. A compact, light bronze looking and black specimen, in fibrous laminæ, exhibiting well its structure. Polished. 6 by 2⅜ ins.

Tyrol.

An analysis of Asbestos from the Tyrol yielded Scheerer :

Silica ...	57·50
Magnesia ...	23·09
Lime	13·42
Oxide of Iron	3·88
Water ...	2·36
	100·35

BERYL GROUP.

No. 687. BERYL, Βήρυλλος, Greek. L., *Beryllus*. Persian name, *Belur*. Aquamarine. A name suggested by Pliny, though not used by him : " *Qui viridilatem puri maris imitantur.*" Emerald. A silicate of alumina and glucina.

Hexagonal. Fracture conchoidal to uneven. Brittle. Hardness = 7·5 to 8·0. Specific gravity = 2·63 to 2·80. Lustre vitreous. Colour pale green to emerald-green passing into blue, yellow, and white. Transparent and translucent. Formula $Be_3Al_2Si_6O_{18}$, or $3BeO . Al_2O_3 . 6SiO_2$, giving the composition :

Silica ...	... ...	67·0
Alumina	... ...	19·0
Glucina	... ...	14·0
		100·0

Another analysis of a Beryl from Siberia yielded Klaproth :

Silica	66·45
Alumina	16·75
Ferrous Oxide ...	0·60
Glucina	15·50
	99·30

A long transparent portion of a crystal of a true aquamarine colour in long striations with parallel lines near the apex, which is well polished naturally. 5 by 1¾ ins.

Mursinka, Ekaterinburg, Siberia.

This specimen formerly belonged to His Imperial Highness the late Duke (Nicholas) of Leuchtenberg, a well-known mineralogist, from whom the author procured it by exchange.

No. 1. Beryl. Aquamarine.

No. 688. PRECIOUS BERYL, of a pale yellowish-green—probably the *Chrysoprasius* of Pliny, and perhaps his *Chryso-lithus* in part. A beautiful transparent coloured specimen. Stratified in structure at base. Polished unfortunately all over. Transparent. 3 by 1⅞ ins.

Ekaterinburg, Siberia.

Coll. H.I.H. the late Grand Duke Nicholas of Leuchtenberg.

No. 689. PRECIOUS BERYL, pale sky-blue, with white, probably the *Aeroides* of Pliny. Translucent. Exhibits iridescence at the top. Cut according to the natural hexagonal shape and polished. 2¼ by 1⅜ ins.

Miask, Ural Mountains.

Coll. His Excellency P. Kotschubey.

No. 690. BERYL, ordinary, in its natural hexagonal form, associated with mica. Opaque. 1⅜ by 1½ ins.

Haddam County, Connecticut, U.S.A.

No. 691. BERYL, Section of, partly cut, pale green. Much cracked on the surface. Exhibits exterior formation. Well polished. 3⅜ by 3½ ins.

Haddam County, Conn., U.S.A.

No. 692. BERYL, Semi-precious. Opaque, pale yellow, hexagonal, crystal associated with mica. Stained black at one part. 3¾ by 2¼ ins.

Haddam County Conn., U.S.A.

Coll. Professor C. U. Shepard.

(a) *Emerald.*

No. 693. EMERALD CRYSTAL (Hexagonal). *Smaragdus* of Pliny. Colour bright emerald-green, due to a small percentage of chromium. Locally known in South America under the name *Canutillos*. In matrix of black Limestone, associated with Quartz.

An analysis of an Emerald from Muso yielded Levy :

Silica	...	...	...	...	67·90
Alumina ...	...	...	...	17·90
Oxide of Chrome, a trace		...		
Glucina	...	...	...	...	12·40
Magnesia ...	...	...	...	0·90
Soda	...	...	...	...	0·70

99·80

With a specific gravity of 2·672.
1⅝ by 1½ ins.

Muso, Santa Fé de Bogota, Columbia.

Pliny describes the Emerald very well and says : " Indeed, there is
no stone, the colour of which is more delightful to the eye ; for
whereas the sight fixes itself with avidity upon the green grass and
the foliage of the trees, we have all the more pleasure in looking
upon the *smaragdus*, there being no green in existence of a more
intense colour than this. And then, besides, of all the precious
stones, this is the only one that feeds the sight without satiating
it. Even when the vision has been fatigued with intently viewing
other objects, it is refreshed by being turned upon this stone ; and
lapidaries know of nothing that is more gratefully soothing to the
eyes, its soft green tints being wonderfully adapted for assuaging
lassitude, when felt in those organs." The Emerald was assigned
the fourth place in the breast-plate of the High Priest, and was
ascribed to Dan.

No. 694. Group of EMERALD CRYSTALS. Pale emerald-
green, embedded in mica schist. With peculiar perpendicular
striations. Translucent to opaque. 2¼ by 1¾ ins.

Ekaterinburg, Siberia.
Coll. of H.I.H. the late Duke (Nicholas) Leuchtenberg.

No. 695. EMERALD. SMARAGD. Hexagonal crystal. Pale
green with mica schist, in which it has been embedded.
1⅞ by 1¼ ins.

River Tokovoya, N. of Ekaterinburg, Siberia.

No. 696. EMERALDS, GROUP OF. SMARAGD. Green transparent crystals embedded in mica schist. 6¼ by 3½ ins.

River Tokovoya, N. of Ekaterinburg, Siberia.
Coll. of His Excellency the late Julian de Siemaschko.

IOLITE GROUP.

No. 1. Iolite or Dichroite.

No. 696A. IOLITE from ἴον = violet, and λίθος - a stone. Dichroite from δίχροος (δίς and χρόα) - two-coloured. Also called Cordierite and Peliom, from πέλιος - smoky blue. A typical specimen of a dark Berlin blue. Translucent, polished on one side. $3\frac{1}{2}$ by $2\frac{3}{4}$ ins.

Bodenmais, Bavaria.

SODALITE GROUP.

Orthosilicates.

SODALITE. Soda and λίθος - a stone. GLAUCOLITH. Isometric. Massive. Fracture conchoidal. Brittle. Hardness = 5·50 to 6·00. Specific gravity of this variety is 2·888. Lavender-blue, blue and sometimes light red. Transparent to translucent.

Formula Na_t (Al Cl) $Al_2Si_3O_{12}$.

Composition of this variety by Hoffman and Rose:

Silica ...	... 38·40
Alumina	... 32·04
Soda ⎰ Potash ⎱	... 24·47
Chlorine	... 7·10
Lime ...	... 0·32
	102·33

Deduct (O = 2Cl) 1·7 = 100. Potassium replaces a small part of the Sodium.

No. 1. Sodalite.

No. 697. SODALITE. A small lavender-blue specimen, with white and brown patches, from the granite-like rock termed *Miascyte*. Polished on both sides. 2¾ by 2½ ins.

Miask, Ilmen Mountains, Urals.

No. 698. SODALITE, lavender-blue specimen, with white and brown patches. A similar specimen to No. 697 from the rock *Miascyte*. Polished. 2⅜ by 2¼ ins.

Miask, Ilmen Mountains, Urals.

No. 699. SODALITE. Lavender-blue, with white and brown patches. A similar specimen, from the rock *Miascyte*. Polished.

Miask, Ilmen Mountains, Urals.

LAZURITE OR LAPIS-LAZULI.

THIS stone can hardly be called a true mineral ; it is an admixture of a colourless and blue substance called Haüyne. It derives its name from the Persian, and means blue colour. It is also called the Armenian stone. Pliny called it *Cyanus* and the Greeks and Romans, *Sapphire*, and, where containing specks of Iron Pyrites, *Sapphirus regilus*. Lapis is, without doubt, the Sapphire of the ancients. Although so well known, its composition has never been strictly determined. It contains some sulphur, about 45 per cent. of silica, 25 per cent. of alumina, with smaller quantities of soda and lime. When reduced to powder it forms the paint ultramarine. It occurs massive, and often contains specs or spots of Pyrites. The finest specimens come from Persia and Lake Baikal. It is also found in Thibet, China, and Chili. It is used extensively for *bijouterie* and for ornaments, as well as for mosaic and inlaying work generally. In Russia the interior of churches are often decorated with it. Many exquisite tables have been also made of the finest Persian lapis-lazuli. The late Emperor William of Germany was an enthusiastic collector of objects in lapis-lazuli, and possessed some of the finest vases known in this material.

LAZURITE. Lapis-Lazuli (from Σάπφειρος, Theophr.). *Sapphiros*, Pliny, 37, 39. Outremer. Native ultramarine.

Isometric—generally massive and compact. Cleavage imperfect. Fracture uneven.

Hardness 5·0 to 5·55. Specific gravity 2·38 to 2·45. Lustre vitreous. Colour a rich azure or Berlin-blue, also violet and greenish-blue. Translucent. Formula Na_1 (NaS_3 Al)

Al$_2$Si$_3$ O, but containing, also in molecular combination, Haüynite, or Sodalite, in varying amount.

Silica ...	31·70
Alumina ...	26·90
Soda ...	27·30
Sulphur ...	16·90
	102·90

Deduct (O = S) 2·9 = 100.

No. 2. Lazurite or Lapis-Lazuli.

No. 700. LAZURITE or LAPIS-LAZULI. A large rectangular specimen of a light-blue colour, the variety called *Asmani*. It is made up of many specimens joined together on a base of slate, associated with white Limestone, and in many places spotted with iron pyrites. 15⅜ by 7⅞ ins.

Andes of the Ovalle, Chili.

No. 701. LAZURITE. Lapis-Lazuli. Dark and light blue. Several specimens joined together, forming an oval containing streaks and spots of iron pyrites. 4 by 2½ ins.

Andes of the Ovalle, Chili.

No. 702. LAZURITE. Lapis-Lazuli. Dark blue ultramarine, with spots of iron pyrites thickly disseminated through it, giving it a mottled appearance. Polished on one side. 3½ by 3½ ins.

Badakshan, Valley of Kokcha, a branch of the Oxus.

No. 703. LAZURITE. Lapis-Lazuli. Light and dark blue block. Ultramarine. Close grained variety, with small specks of iron pyrites. Well polished. 4 by 2¾ ins.

Persia.

No. 703A. Ball of LAZURITE. Lapis-Lazuli. Light blue with white spots. Diameter 1¾ ins.

Cut, polished and drilled at Oberstein, but from the Andes of the Ovalle, Chili.

GARNET GROUP.

No. 1. Garnet.

No. 704. GARNET ("Ανθραξ, pt. Theophr.), from the Latin *Granitus*, grain-like. Almandite, precious or Oriental Garnet. Silicate of alumina and iron associated with Smaragdite. Hardness 6·5 to 7·5. Specific gravity 3·15 to 4·30.

Composition : Formula 3FeO. Al_2O_3. $3SiO_2$, equal to

Silica...	... 36·20
Alumina	... 20·50
Ferrous Oxide	... 43·30
	100·00

Ferric iron replaces the aluminium to a greater or less extent. A thin oval specimen associated with Smaragdite. Very beautiful by transparent light. Polished. 3¼ by 2¼ ins.

Orawitza, Hungary.

Coll. of Staatsrath Braun of Vienna.

(a) *Carbuncle.*

No. 705. Ball cut out of a solid Carbuncle (Latin *Carbunculus*, from *Carbo*, coal). Exhibiting an asteroid or star of four lines, particularly brilliant in the sun or by artificial light. Colour carbuncle-red. Extremely rare. Cut and polished in England. Diameter 2½ ins.

River-beds of Ratnapoora, Ceylon.

The third precious stone in the breast-plate of the High Priest was the Carbuncle, and was ascribed to Judah. It has always been symbolical of Majesty or Royal dignity.

TOPAZ GROUP.

TOPAZ. Not Τοπάζιος, Gk. ; nor the Topazos of Pliny or Agricola. *Chrysolithos*, pt. Pliny 37, 42. *Pitdah*, Hebrew. Orthorhombic. Fracture subconchoidal to uneven. Brittle. Hardness 8·0. Specific gravity 3·40 to 3·65. Bluish-yellow, white, etc.

Composition : Fluosilicate of Alumina. Formula (Al (O, F_2),), Al SiO_4, Groth. ; equal to

Silicon	... ...	15·50
Aluminium	... ...	29·90
Fluorine	... ...	17·60
Oxygen	... ...	36·90
		99·90

Cut extensively as gems. Was assigned the second position in the breast-plate of the High Priest, and had engraved on it the name of Simeon, the second son of Jacob. Professor Ruskin, in his Lecture on the "Symbolic use of Precious Stones in Heraldry," describes the topaz as being "symbolic of the sun, like a strong man running his race rejoicing, standing between light and darkness, and representing all good work." The Topaz was included by Pliny and earlier writers, as well as by many later ones, under the name *Chrysolite*.

No. 1. *Topaz.*

No. 706. TOPAZ CRYSTAL, beautifully terminated with peculiar indentations or markings upon the table, caused, probably, by the pressure of other minerals which had become disintegrated. Pale blue. Well crystallized, and transparent to translucent. A most perfect and valuable specimen. Associated with a small piece of black Tourmaline. 2¼ by 2 ins.

Alabashka, Mursinsk, Perm, Urals.

Coll. H.I.H. the late Duke Nicholas of Leuchtenberg.

No. 707. TOPAZES, group of crystals, many doubly terminated. Amber coloured, of two tints. Associated with black crystals of Quartz. 3 by 3 ins.

Adun-Tschilon, Nertschink, Transbaikal.

No. 708. TOPAZ, rough bluish white. Translucent, waterworn pebble. This variety is often term *gouttes d'eau*, owing to its limpidity. Exhibits iridescence 1⅜ by 1¼ ins.

From the River-beds of the Gulgong District, New South Wales.

No. 709. TOPAZ, rough, water-worn. Translucent. Exhibits iridescence and fracture. Bluish-white. 1¼ by 1⅛ ins.

River-beds of the Gulgong District, New South Wales.

No. 710. TOPAZ, rough, water-worn. Translucent. Bluish-white. Exhibits fracture and iridescence. 1¼ by 1 ins.

River-beds of the Gulgong District, New South Wales.

These rough Topazes, when perfectly transparent and cut into gems, are like brilliants, and exhibit a great amount of fire.

EPIDOTE GROUP.

Thulite, a Variety of Zoisite.

THULITE, generally regarded as an Epidote until the research of Professor A. Descloiseaux proved it to be a rose-red variety of Zoisite, a silicate of alumina, lime, and iron, with a trace of magnesia and a little iron. Specific gravity 3·124. Hardness 6·0 to 6·5.

An analysis by Gmelin yielded :

Silica	...	42·81
Alumina ...	...	31·14
Ferric Oxide ...	...	2·29
Manganese Sexquioxide ...		1·63
Lime...	...	18·73
Soda...	...	1·89
Water	...	0·64
		99·13

No. 1. Thulite.

No. 711. THULITE, after Thule, a name by which the Scandinavian Peninsula was known to some of the ancient geographers. A very fine large rose-red specimen, associated with Quartz. Occurs in small lamellar masses, easily cleavable in one direction, or in semi-crystals with canal-like cavities running in their longest direction. Well polished. 12½ by 10½ ins.

Kleppau, Parish of Souland, Tellemarken, Norway.

A most striking and beautifully coloured specimen. Now largely introduced as a decorative mineral, and by reason of its beautiful soft colour is certain to be much sought after.

No. 712. THULITE, a deep rose-red coloured specimen, associated with Quartz. Polished. 8½ by 5¾ ins.

Kleppau, Parish of Souland, Tellemarken, Norway.

No. 713. THULITE, a variety of Zoisite. A deep rose-red colour, associated with Quartz. Well polished. 5 by 4 ins.

Kleppau, Souland, Tellemarken, Norway.

No. 714. Ball cut out of THULITE, of a bright light rose-red colour, with white patches. Diameter 1¾ ins. Cut and polished on the Marne.

Kleppau, Tellemarken, Norway.

No. 715. Ball cut out of THULITE of a deep bright rose colour, with black markings. Diameter 1½ ins.

Kleppau, Tellemarken, Norway.

PREHNITE.

An Orthosilicate not included in any foregoing group.

PREHNITE, named by Werner after Col. Prehn, in 1790, who first discovered it at the Cape of Good Hope. A hydrous silicate of alumina and lime. Hardness 6·0 to 6·5. Specific gravity 2·80 to 2·95.

No. 1. Prehnite.

No. 716. PREHNITE. A peculiar mottled grey white specimen. Looking like Jade for which it is sometimes mistaken, associated with white mica. Formula : $H_2Ca_2Al_2Si_3O_{12}$ equal to

Silica	...	43·70
Alumina	...	24·80
Lime...	...	27·10
Water	...	4·40
		100·00

$7\frac{1}{8}$ to $3\frac{3}{4}$ ins.

Cradock, South Africa.

B. HYDROUS SILICATES.

MICA GROUP.

AGALMATOLITE, from ἄγαλμα = an ornament, a thing to glory in, the image of a god. *Bildstein* of China; also Pagodite, from pagoda, the Chinese carving it into miniature pagodas, images, etc. A variety of Pinite, a hydrous silicate of aluminium and potassium, but containing more silica. Hardness 2·5 to 3·5. Specific gravity 2·6 to 2·85.

No. 1. Agalmatolite or Pagodite.

No. 717. AGALMATOLITE. Small pea-greenish and yellow specimen occurring in spherules. An analysis of the pea-green variety from China yielded Schneider:

Silica,	... ...	... 63·29
Magnesia	... ...	... 31·92
Oxide of Iron	...	... 2·27
Oxide of Manganese		... 0·23
Alumina	... ...	... 0·53
Water	... ...	... 0·78
		91·02

Polished. 1¾ by 1¼ ins.

Mooi River, Transvaal, South Africa.

No. 2. Lepidolite or Lithia Mica.

LEPIDOLITE. Lithia Mica. Schuppenstein, *Germ.* A variety of Mica, from λεπίς a scale, and λίθος=a stone. Violet grey, or lilac. A silicate of alumina, potassium, and lithia, with oxide of manganese and fluorine. Rhombic. Hardness 2·5 to 4·0. Specific gravity 2·8 to 2·9.

Q

No. 718. LEPIDOLITE. Lithia Mica. A lilac-coloured specimen of two tints. The scaly micaceous structure very distinctly exhibited. An analysis of this Moravian Lepidolite yielded Regnault:

Silica	52·40
Alumina	26·80
Manganese Sesquioxide	1·66
Potassium	9·14
Lithia	4·85
Fluor Spar	4·18
	99·03

Polished on one side. 8 by 6 ins.

Rozena, Moravia.

Used for inlaying, its uncommon and subdued colour being of great value for softening more pronounced colours.

No. 719. LEPIDOLITE. Lithia Mica. Lilac-coloured specimen of two tints. Exhibiting its micaceous structure clearly. Polished. 7¾ by 6 ins.

Rozena, Moravia.

SERPENTINE GROUP.

Serpentine.

SERPENTINE. 'Οφίτης, *pt. Dioscor.*, 5,161, from the Greek ὄφις = a serpent. Marmor Serpentinum. Ophite. Monoclinic. Usually massive. Colour various greens, brownish-red and yellow, never bright. Fracture conchoidal or splintery. Smooth or greasy feel. Translucent to opaque. Hardness 2·5 to 4·0 owing to impurities, and rarely 5·5. Specific gravity 2·50 to 2·65.

Composition: A magnesian silicate. $H_4Mg_3Si_2O_9$, or $3MgO$. $2SiO_2$. $2H_2O$, equal to

Silica...	...	44·10
Magnesia	...	43·00
Water	...	12·90
		100·00

No. 1. *Serpentine.*

No. 720. SERPENTINE, rectangular. Yellowish-green, with dark patches. Mottled in appearance. Polished. 5 by 2¼ ins.

New Jersey, U.S.A.

An analysis of this variety of Serpentine yielded Garrett:

Silica... ...	42·32
Magnesia ...	42·23
Ferrous Oxide	1·28
Alumina ...	0·66
Water ...	13·10
	100·29

No. 721. SERPENTINE, rectangular. Yellowish-green with dark patches. Mottled in appearance. Polished. 5 by 2¾ ins.

New Jersey, U.S.A.

No. 722. SERPENTINE, sometimes called Precious Serpentine, bright oily-yellow. A hydrous silicate of magnesia with a small percentage of alumina and ferrous oxide. An analysis by Helland* of the Snarum Serpentine yielded :

Dioxide of Silicon ...	...	42·72
Ferric Oxide ...	...	0·06
Ferrous Oxide ...	...	2·25
Magnesia	...	42·52
Water	...	13·39
		100·94

Hardness 3·0. Specific gravity of this variety 2·53. Well polished. 7¾ by 5 ins.

Snarum, Norway.

A freshly discovered deposit of this beautiful and striking coloured Serpentine was lately discovered in Norway, and by the assistance and encouragement of the King is being largely developed and made into ornaments, such as vases, boxes, tazze, etc.

No. 723. SERPENTINE, sometimes called Precious Serpentine, a bright oily-yellow. Well polished. 5 by 5 ins.

Snarum, Norway.

Massive (a) *Bowenite.*

No. 724. BOWENITE (Nephrite of *Bowen*). A large slab of a greenish-white colour, with a very fine granular texture, closely resembling Jade, for which it is often mistaken. It has the unusual hardness for a Serpentine of 5·5 to 6·0, and a specific gravity of 2·594 to 2·787. Polished on one side. 14 by 6¾ ins.

Smithfield, Rhode Island.

This is the mineral termed *Sang-i-yashm* in Persian. Ac-

* "*Pogg. Ann.*," 148, 329, 1873.

cording to Forbes' and Platts' dictionaries, a Jasper, especially from China supposed to be an infallible charm against lightning.* An analysis made by G. T. Prior, M.A., in the British Museum Laboratory yielded:

Silica	44·73
Magnesia ...	42·64
Alumina ...	0·32
Ferrous Oxide	0·33
Lime, a trace ...	
Water	12·21
	100·23

Showing that it is in reality a Serpentine, the composition of Serpentine being broadly:

Magnesia ...	...	...	43·48
Silica	...	...	43·48
Water	...	...	13·04
			100·00

This mineral is used at Bhera, in the Shahpur district of the Punjaub, for the manufacture of small ornamental articles.

Lamellar (b) *Williamsite.*

No. 725. WILLIAMSITE. A lamellar variety of impure Serpentine, graduating into a massive granular variety described by Prof. C. U. Shepard of Amherst College. Apple-green colour. The analysis of this mineral by Smith and Brush yielded:

Silica	42·10
Magnesia ...	41·50
Ferrous Oxide...	2·43
Oxide of Nickel	0·45
Water	12·70
	99·18

Translucent. Hardness 4·5. Specific gravity 2·59 to 2·64. 5½ by 3¾ ins.

Texas.

* *Notes on Bowenite or Pseudo-Jade from Afghanistan by Major-General C. A. McMahon, "Min. Mag.," Vol. ix., 187, 1890.*

Fibrous (c) *Chrysotile.*

No. 726. CHRYSOTILE, a delicate fibrous mineral, constituting seams in Serpentine. It includes, Dana says, most of the silky *amianthus* of Serpentine rocks, and much of what is popularly called Asbestos. An analysis of this mineral by Brush from Newhaven yielded :

Silica	44·05
Magnesia	39·24
Ferrous Oxide... ...	2·53
Water	13·49
	99·31

Specific gravity 2·219. A small olive-green silky fibrous specimen running in irregular strata through Serpentine. Polished on one side. 2¾ by 2½ ins.

Newhaven, Connecticut, U.S.A.

SULPHATES.

Barite Group.

BARITE, from βάρος = weight, or βαρύς = heavy. Sulphate of Baryta. "Cawk" of the Derbyshire miners. Heavy spar. Orthorhombic. Fracture uneven. Brittle. Hardness 2·50 to 3·50. Specific gravity 4·30 to 4·60. Lustreous, vitreous, inclining to resinous. Colour white, inclining to yellow; also brown grey, blue, and red. Becomes fetid sometimes when rubbed Translucent to opaque.

Composition : Barium Sulphate. Ba SO₄ equal to :

Sulphur trioxide	... 34·3
Baryta... ...	... 65·7
	100·0

Strontian and calcium sulphate are sometimes present.

No. 1. *Barite.*

No. 727. BARITE. Sulphate of Baryta. Heavy spar. Opaque. A large irregular circular white section, principally of Barytes, associated with Galena (sulphide of lead) with Calcite (carbonate of lime) and transparent white Fluor Spar. Semicrystallized, surrounded by a thin layer of Galena and opaque Calcite. Polished on one side. 7¾ by 6 ins.

Matlock district, Derbyshire.

A most interesting specimen, exhibiting well the association of metallic and non-metallic minerals.

No. 728. BARITE. Sulphate of Baryta. Stalactitic. Cross section of a stalactite exhibiting concentric deposition similar to calcareous stalactites. Brown. Polished on one side. 1⅞ by 1¾ ins.

Matlock, Derbyshire.

A pretty specimen, often mistaken for " petrified wood."

No. 729. GYPSUM. Γύψος—the ancient name for the mineral, from γῆ = the earth, and ἔψειν = to boil, or smelt.   Sulphate of lime.   A fibrous satin-spar variety, having the pearly opalescence of Moonstone (Σεληνίτης).   Monoclinic.   Hardness 1·5 to 2·0.   Specific gravity 2·314 to 2·328 when in pure crystals.

Hydrous calcium sulphate, $CaSO^4 + 2H_2O =$

Sulphur Trixoxide	...	...	...	46·6
Lime ...	...	...	...	32·5
Water...	...	...	...	20·9
				100·0

Lapis specularis of Pliny.   Cut into the shape of an egg, exhibiting an opalescent light when turned.   Length, 2½ by 1¾ ins.

Matlock, Derbyshire.

No. 2.   Gypsum.

No. 730. GYPSUM or Sulphate of Lime.   Cut into the shape of an egg, exhibiting an opalescent light, caused by its fibrous structure   Length, 2½ by 1¾ ins.

Matlock, Derbyshire.

HYDROCARBON COMPOUNDS.

Oxygenated Hydrocarbons.

AMBER OR SUCCINITE.

THIS well known mineral has attracted particular attention in all times through its electrical properties, and is mentioned frequently by Homer as Electron ($\mathring{\eta}\lambda\epsilon\kappa\tau\rho\text{o}\nu$)—who refers to it in his description of the Trojan War, mentioning that the women wore necklaces of Amber. Theophrastus, Dioscorides, and Pliny, all knew it by the same name, Electrum, whence has been derived the word electricity. Thales was greatly taken back at its electrical properties, and thought there was a soul in the Amber, but Tacitus, in his "Germania," states it is a resin exuded by certain coniferæ. Pliny mentions, says Dana, as one proposed derivation of *Electrum* the following fable : The sisters of Phaëthon, the son of Apollo, became changed into poplars, which every year shed their tears on the bands of the Eridanus or (Padus). To these tears was given the name *Electrum*, from the fact that the sun was usually called elector. Another fable has it that it comes from *Electrides*, the name of certain islands in the Adriatic : or another, from *Electrides*, the name of certain stones in Britannia, from which it exudes. He gives it as his opinion, continues Dana, that "Amber is an exudation from trees of the pine family, like gum from the cherry, and resin from the ordinary pine " ; and as a proof that it was once a liquid, alludes to the gnats, etc., in it, observing also that it had long been known as *Succinum* because of this origin : " *Quod arboris succum prisci nostri credidere.*" He says that in his time it was " in request among women only." But it had been so highly valued as an object of luxury that a very diminutive human effigy made of amber had been known to sell at a higher price than living men, even in stout and vigorous health.

It will be thus seen that Pliny was correct in his views, for it is now fully admitted that Amber is truly of vegetable origin, altered by fossilization as well as many other fossil resins ; the inference being derived from its proximity to coal or fossil wood with the occurrence of insects embedded in it. Many of these show evidence of death struggles after being entangled in the then viscuous fluid, a leg or wing being found detached some distance from the body—lost in the struggle to escape entombment. Philemon called it a fossil, and says it is found in two different localities in Scythia, but Tacitus, judging most probably from the insects it often contains, concluded, and correctly also, that it was a vegetable juice, and called it *Succinum.* Theophrastus has stated, Pliny remarks, that Amber is extracted from the earth in Liguria,* and that Demonstratus calls Amber *Lyncurion,* and he says that it originates in the urine of the wild beast known as the lynx.

It was known to and imported by the Phœnicians, who sailed to the Glessarian Islands solely to procure it. By some of the ancients it was called *Lyncurium,* though both the *Zircon* or *Jacut* of Pliny, and Tourmaline also were known by the same name, both gems having strong electrical properties. Many fossil resins are termed Amber which are not so in the true sense of the word, Amber being, according to the great authority upon fossil resins (Dr Otto Helm, of Dantzig), distinguished by containing succinic acid and being derived solely from the *Pinus succinifer.* This view is sustained by Herr Conwentz (" Monograph der Baltischen Bernsteinbaume," Dantzig, 1890). Göppert has shown† that at least eight species of plants as well as the *Pinites succinifer* have afforded these fossilized resins, and Dana states he enumerates 163 species as represented by remains in them. Besides the true Pine species belonging to the family, Abietineæ and Cupressineæ have probably contributed to them.

---

* *Amber is found in St. Paulet in the Department du Gard, and at Aix in the Department of the Bouches du Rhone, which are not very distant from the territory of Liguria.*

† " *Ber. Ak. Berlin,*" 450, 1853, " *Am. J. Sc.*" 18, 287, 1854.

A resin found plentifully in Burmah has been termed for centuries Amber, and many accounts of the Amber mines in the Hukông Valley in the Patkoi range* have been published. Dr. Griffiths describes the Amber mines in his "Journal of Travels in Assam, Burmah," 1847, giving a description of the pits above the plain of Meinkhon ; and latterly Dr. Fritz Noetling has described them in the "Geological Survey of India," January 1893, but not under the former name of Amber, but under that of Burmite, a new fossil resin, from Upper Burmah, the name being suggested by Dr. Helm. None of the Burmese so-called Amber contains succinic acid hence its separation from Amber proper.

The Prussian coast of the Baltic yields the greatest quantity of Amber, from Dantzig to Memel, especially between Pillau and Dorf Gross-Hubnicken, where it is mined in a systematic manner as well as being thrown up on the shores by the waves after a storm.   It occurs in a bed of bituminous coal, which extends under the sea and is washed out during the autumnal storms and carried by the waves to the shore. Pliny was aware of it being thrown up by the sea, as he says: "It is thrown up upon the coasts, in so light and voluble a form that in the shallows it has all the appearance of hanging suspended in the water." On the land a shaft is sunk through the coal-bed more than 100 feet deep. In 1891 out of 2,430,000 cubic feet of "blue earth" were taken 405,000 pounds of amber valued at £100,000. It is also found in England, many specimens being thrown up at Reculvers and Westgate, as well as on the coasts of Norfolk, Essex, and Suffolk ; also on the Danish coast, and largely on the Russian Baltic and Westphalian coasts.   The resins resembling Amber occur in France, Poland, Norway, Switzerland, and other European localities, as well as Upper Burmah, Mexico, and other countries.

Large specimens of Amber have been found. One exhibited at the Royal Museum in Berlin weighs 18lbs. Another

* "*Journal of the Asiatic Society of Bengal,*" *Vol.* vi., *April,* 1837.

at Ava in India is nearly as large as a child's head, and weighs two and a half pounds, and is intersected by veins of carbonate of lime, from a line to one-twentieth of an inch. The specimen No. 732 in this collection is unusually large, measuring 8 inches by 6. A sailor is reported to have discovered a large specimen in an unusual manner. Seating himself on a rock for some time, he found on attempting to rise, he could not, the natural heat of his body having attracted the Amber upon which he had inadvertently sat, and for a short time held him prisoner.

Many of the *tumuli* of the Neolithic period in Denmark, when opened contain necklaces of Amber, which were evidently buried with the bodies. At the mines it is assorted in different qualities, and receives technical names. These are :

1. The *Exquisite* specimens—pure, transparent, and compact.

2. The *Ton* stones, weighing up to four ounces—the finest being used for jewelry purposes, and the impure crushed for medicine.

3. *Nodules*, which are smaller.

4. *Varnish* stones still smaller, but very pure and hard, so as to be pulverized easily for use in varnishes, and for making sealing-wax.

5. *Sandstones.* Small opaque and perforated specimens.

6. Refuse, which is mixed with gum-arabic, shellac and pastes.

It is extensively used for *bijouterie*, taking a high polish ; being made into beads, bracelets, ear-rings, buttons, etc. Pliny mentions it as being in request by the women only. Now it is the men who create the demand for it, and very largely for mouth-pieces to pipes and cigar-holders, associated with meerschaum. It is polished generally on a leaden wheel with pumice stone, then with linen and rotten-stone, and finally by the hand.

It is very attractive, by often containing insects, such as flies and beetles, and sometimes, but very rarely, small fish. The great demand for specimens of Amber containing flies soon created a supply of " adulterated " specimens. This was easily effected by melting the Amber, throwing in the insects, and letting it cool ; but, like all such imitations, the makers, taking the first insects they could place their hands on, revealed their secret to the naturalist by enclosing species of insects and flies often not known in the countries from which the various specimens of Amber were obtained. Many imitation specimens of Amber with insects have been made at Oberstein-on-the-Nahe.

The most extraordinary collection of Amber specimens is naturally to be found at Dantzig. In the Quirinal at Rome the walls of a small boudoir are made entirely of a cloudy Amber, the effect being, perhaps, more *bizarre* than beautiful.

Certain classes of the Austrian, Bulgarian, and Turkish women are very superstitious about Amber jewelry, and wear it, believing " it ensures the wearer good luck, and a long enjoyment of the qualities that make them attractive."

SUCCINITE. AMBER. Ἤλεκτρον, Homer, etc. *Electrum. Lyncurium. Bernstein,* Germ. ; *Ambre,* Fr. Occurs in irregular masses, no cleavage ; fracture conchoidal. Optically anisotropic, exhibiting under the polariscope bright interference colours. Hardness 2·0 to 2·5. Specific gravity 1·050 to 1·096, *Helm. ;* 1·061 to 1·112 *Damour.* Lustre resinous. Melts at 250° to 300° ; softens at 150°. Colour yellow, brownish, and reddish—clouded. Negatively electrified on friction. Transparent to translucent.

Composition : Ratio for C, H, O, equal to 40 ; 64 ; 4 ; equal to :

Carbon	78·94
Hydrogen	10·53
Oxygen	10·53
	100·00

No. 1. *Amber.*

No 731. AMBER. SUCCINITE. *Electrum.* A round nodular specimen of a most beautiful rich yellow and brown colour. Translucent to opaque. Polished all over. 4 by 2½ ins.

Prussian coast of the Baltic·

A very beautiful and characteristic specimen, both for colour and shape.

No. 732. AMBER. SUCCINITE. *Electrum.* A large irregular shaped specimen of a rich light and dark yellow with a brownish colour. Translucent. Well polished all over. 8 by 6 ins.

Prussian coast of the Baltic.

An unusually large and magnificent specimen.

No. 733. AMBER. SUCCINITE. Enclosing a large quantity of insects with their carapaces. Of a pale yellow colour. Beautifully transparent. Well polished. 5 by 4½ ins.

Prussian coast of the Baltic.

No. 734. AMBER. SUCCINITE. With a yellow coloured interior and a reddish exterior. Opaque. Polished on one side. 3¾ by 2½ ins.

Forest-bed of Cromer, Norfolk.

No. 735. AMBER. SUCCINITE. Yellow interior, with reddish exterior. Similar to No. 734, being half of the same stone. Opaque. Polished on one side. 5 by 4½ ins.

Forest-bed of Cromer, Norfolk.

No. 736. AMBER. SUCCINITE. A reddish specimen, transparent, with a white translucent cloud; and a deposition, layer-like, of white amber. One side polished. 2 by 1¾ ins.

Prussian shores of the Baltic.

No. 737. AMBER. SUCCINITE. Cut *en cabochon,* enclosing insects. Yellow and perfectly transparent. Polished all over. 1⅞ by 1⅛ ins.

Shores of the Baltic.

No. 738. AMBER. SUCCINITE. In shape of a pillar. Enclosing a spider with a small fly in the corner, which it was most probably chasing when they both became entombed. Transparent. Bevelled edges. 1⅛ by ⅜ ins.

Shores of the Baltic.

No. 739. YELLOW AMBER. SUCCINITE. Cut *en cabochon.* Enclosing an insect (Hymenoptera). 1¼ by ¾ in.

Baltic Ocean.

No. 740. AMBER. SUCCINITE. Pale yellow. Cut into the shape of a boot, enclosing a spider. Perfectly transparent and polished all over. 1½ by ½ in.

Baltic shores.

No. 741. AMBER. SUCCINITE. Cut *en cabachon.* With flaw in centre. Containing a spider. 1 by ¾ in.

Baltic shores.

No. 742. AMBER. SUCCINITE. Triangular. Cut for a seal, to turn on a swivel. Pale yellow. Containing an insect. Transparent. 11/16 by 11/16 in.

Shores of the Baltic.

No. 743. AMBER. SUCCINITE. Drop. Pale yellow. Enclosing an insect. Transparent. Polished all over. 1¼ by ½ in.

Baltic.

No. 744. AMBER. SUCCINITE. *En cabochon.* Containing a large insect. Perfectly transparent and polished all over. ¾ in. diameter.

Shores of the Baltic.

No. 745. AMBER. SUCCINITE. Pale yellow. Oval. Clouded in centre. Polished all over. 1¾ by 1¼ ins.

Shores of the Baltic.

No. 746. AMBER. SUCCINITE. Cut *en cabochon.* Enclosing an insect. Transparent. ½ in. diameter.

Shores of the Baltic.

No. 747. AMBER. SUCCINITE. Cut *en cabochon.* Containing a red looking insect. Transparent. ½ in. diameter.

Shores of the Baltic.

No. 748. AMBER. SUCCINITE. Cut *en cabochon* and round.
Pale. Containing a small, very pretty beetle, with iridescent
back. ⅜ in. diameter.

Shores of the Baltic.

No. 749. AMBER. SUCCINITE. Cut *en cabochon,* round.
Containing an insect. Transparent. ⅜ in.

Shores of the Baltic.

No. 750. AMBER. SUCCINITE. Of a deep *reddish brown*
by transmitted light. With top and bottom surface polished.
1¾ by 1⅜ ins.

Shores of the Baltic.

No. 751. AMBER. SUCCINITE. Light transparent. Pale
yellow. Cut into a seal handle, and enclosing part of the
remains of an insect. Transparent. 2¾ by 1 in.

Dantzic.

No. 752. AMBER. SUCCINITE. Pale yellow. Containing
an insect. Oblong, with *cabochon* sides to a flat top. Transparent, and well polished. 1½ by 1 in.

Shores of the Baltic.

No. 753. AMBER. SUCCINITE. Containing insect. *Cabochon*
cut specimen. Pale yellow. Transparent. 1½ by ¾ in.

Shores of the Baltic.

No. 754. AMBER. SUCCINITE. Triangular specimen, with corners and sides bevelled. Of a pale transparent yellow, containing an insect. Polished all over. $1\frac{1}{4}$ by $1\frac{1}{8}$ ins.

Shores of the Baltic.

No. 755. AMBER. SUCCINITE. Pale transparent drop. Polished all over. $1\frac{1}{8}$ in.

Shores of the Baltic.

No. 756. AMBER. SUCCINITE. Pure, transparent specimen. A fracture in the interior looks like the carapace of an insect. $2\frac{1}{4}$ by $1\frac{1}{2}$ ins.

Shores of the Baltic.

R

METEORITES.

UNDER the name Meteorites are classified the three bodies known as Meteoric Irons, or Aerosiderites, Meteoric Iron and Stone combined (called Siderolites), and Meteoric Stones (known as Aerolites). That these remarkable bodies were known to the ancients, and greatly venerated, there is no doubt. It is mentioned in the Scriptures that after the Battle of Gibeon, and during the flight of the Canaanites, great stones were cast down from Heaven, so that people were slain by them ; but the wording is ambiguous, and might apply to large hailstones. One famous stone Mr. L. Fletcher, keeper of the minerals in the British Museum, in his introduction to the "Study of Meteorites," mentions as falling in Phrygia, where it was preserved for many generations, and worshipped at Pessinus by the Phrygians and Phœnecians as Cybele, "the Mother of the Gods," subsequently being taken to Rome by King Attalus, about 204 B.C., with great ceremony. Other early Meteorites are recited by Plutarch as falling in Thrace, 470 B.C. A shower, Livy says in his " History of Rome," took place about 652 B.C. ; whilst they are mentioned in Chinese manuscripts 644 B.C. Pliny also refers to stones which fell in Asia Minor and Macedonia.

The first truly authenticated fall of a Meteorite is that which fell at Ensisheim, in Elsass, on the 7th November, 1492, since which date continual falls have and are taking place yearly, scattered indiscriminately over all the months and without reference to latitudes or seasons, many falling in perfectly calm weather, as well as during thunderstorms. One in England, of great celebrity, fell near a labourer, whilst working at Wold Cottage, Scarborough, in Yorkshire, weighing 56 lbs., which is now in the British Museum, South Kensington. Another celebrated meteoric stone fall is that of l'Aigle, in the Orne Department, France. A fire ball was seen on Tuesday, April 26th, 1803, about 1 p.m., a few moments after which a terrific explosion occurred, lasting over five minutes, and being heard for seventy-five miles round. Before the explosion a fire-ball or meteor was observed at adjoining towns, but not at l'Aigle itself, and on the same day stones, estimated to number between two and three thousand, supposed to be fragments of this ball, fell in the neighbourhood of l'Aigle. An exhaustive report was drawn up by M. Biot, of the French Academy, which left no doubt as to the origin of the fall, and that Meteorites really came from outer space. In more recent times a great fall took place at Pultusk, in Poland, on January 30th, 1868, when several thousands of stones fell, from the size of a nut to that of a man's head (one in the British Museum weighs 14,587·0 grammes), whilst another large fall took place at Möcs, Kolos, in Transylvania, on February 3rd, 1882, the stones being scattered over many villages, one of which, with a complete carbonaceous crust is in the "Derby" Collection, No. 764. Out of the large number of meteoric falls known (about 400 representing the three classes), it is peculiar that most of the Aerolites or Stony Meteorites *have been seen* to fall, or their dates have been well authenticated, whilst of the siderites or irons, only nine have been seen to fall out of about 150. The largest stone known fell at Knyahinya, Hungary, on June 9th, 1866, and weighs 647 lbs. It is in the Vienna Museum. The largest metoric iron is that found in the Province of the Tucuman, in the Argentine Republic, by Indians, who took it to be an iron mine. It was estimated to weigh, when discovered, 30,000 lbs. A large block, 1,400 lbs. in weight of this specimen is in the National collection at the British Museum, South Kensington. Meteoric iron

R 2

differs essentially from any metallic substance found on this earth, by being malleable in its natural state, knives and even sword-blades having been made of it. A very fine paper-knife, the blade of which is about eight inches long, made from the Butler Meteoric Iron Bates Co., is in the possession of Mr. Alfred Morrison, its locality being beautifully etched upon it.

The great question of the origin of these remarkable bodies has undergone continual change, from the belief that they were simply common stones struck by lightning, or were ejected by volcanoes, to the more advanced, and most probably correct solution, that they really constitute what are known to us as comets—in fact, that a comet is really a mass of Meteorites, and that what we term its tail are the gases driven from them; the luminosity of the comet being caused partly by reflected sunlight, and the combustion of the Meteorites by the sun's rays. The whole question as it now stands is lucidly explained by Mr. L. Fletcher,* of the British Museum.

The late Earl of Derby purchased the following Meteorites to illustrate the three classes into which these remarkable and deeply interesting bodies are divided.

METEORIC IRONS, AEROSIDERITES OR SIDERITES.

Native Iron Masses, generally Nickeliferous, and containing Schreibersite, Troilite, Graphite, etc.

No. 757. METEORIC IRON. Weight 127 grammes. Fell at Glorieta Mountain, one mile north-east of Canoncito, Santa Fé County, New Mexico, U.S.A.

Portion of the original specimen, cut and well polished. 3¼ by 3 ins.

> This Siderite was found in the year 1884, and was described by the American mineralogist Kunz, in the "American Journal of Science," 1885 (ser 3, vol. xxx., page 235)

---

* *"Introduction to the Study of Meteorites,"* by *L. Fletcher, M.A. (British Museum).*

No. 758. SIDERITE or METEORIC IRON, slice of, commonly called the "Butler Iron." Turned up by a plough, a long time after which it came to the knowledge of Broadhead, who mentioned it in 1875. "American Journal of Science," 1875, ser. 3, vol. x., page 401. Weight 137 7 grammes.

On the polished surface will be noticed a nodule of a tomback-brown colour, which goes entirely through the stone. This mineral is Troilite (Pyrrhotite pt.), a sulphide of iron, usually accepted as FeS yielding—

Sulphur	...	...	36·4
Iron	...	...	63·6
			100·0

but M. Meunier, of the School of Mines in Paris, regards it as identical with Pyrrhotite, another sulphide of iron often containing nickel.

It has a hardness of 4·0, with a specific gravity 4·75 to 4·82. It is pretty common in nodules in Meteorites disseminated more or less sparingly through the mass, occurring also, Dana says, in thin veins, "usually separated from the iron by a thin layer of Graphite." It was named after Dominico Troili, who in 1766 discovered it in a Meteorite that fell at Albarato, in Modena, and described it. Polished on surface. 2½ by 1⅜ins.

Butler, Bates County, Missouri, U.S.A.

SIDEROLITES, OR LITHOSIDERITES.

Meteoric Iron and Stone Combined.

Nos. 759, 760, 761, 762. SIDEROLITES or METEORIC IRONS and STONE combined. Four specimens, weighing respectively No. 759, 14·8 grms.; No. 760, 6·6 grms.; No. 761 9·6 grms.; No. 762, 5·7, grms. Called the "Estherville Siderolite," having fallen at Estherville, Emmet County, Iowa, U.S.A., on May 10th, 1879. A very large specimen is in the National (British Museum) Collection, weighing 116·487 grammes.

AEROLITES OR LITHOLITES.

Meteoric Stones.

No. 763. AEROLITE or METEORIC STONE, a slice cut from what is generally called the "Amana" meteor. Weight 31·7 grammes. Fell on Feb. 12, 1875. 2½ by 2¼ ins.

> Amana, West Liberty, Iowa county, Iowa, United States, America.

No. 764. AEROLITE or METEORIC STONE, with a complete carbonaceous crust. Weight 25 grammes. Fell on February 3rd, 1882, at Mocs (Visa), Kolos, Transylvania. 1⅞ ins.

An Ungrouped Metallic Specimen.

No. 765. SLICKENSIDE of Iron Pyrites. 6 by 5¾ ins.

> Alston District, Cumberland.

Slickenside is a name generally given to *specular* Galena, a sulphide of lead, or a variety with a polished face, very often with indentations or cavities, which are also polished naturally in the interior. This polish was thought to be due to a process of rubbing one portion of a stratum against another, or by a slipping of rocks, but it is more probably caused by "electricity." When miners strike the slickensides of lead, in Cumberland, they often explode with a dangerous violence. This specimen is a hard variety not often met with, and has a naturally polished surface, with a thin coating of iron pyrites of a brassy golden colour.

ROCKS.

No. 766. EUPHOTIDE (Gabbro in part), a metamorphic rock, of a Smaragditic variety. Consists of whitish Saussurite, mixed with emerald-green Smaragdite. Very tough, having a specific gravity of 2·9 to 3·4. The *Verde di Corsica* of decorative art. Well polished. 5¾ by 4½ ins.

Orezza Valley, Corsica.

A beautiful and most instructive specimen, illustrative of the varieties used for making table-tops.

No. 767. PORPHYRITE, consisting of crystals of pale green Feldspar embedded in a diabase ground mass. Well polished. 4¾ by 3½ ins.

Orezza Valley, Corsica.

No. 768. PORPHYRITE, consisting of crystals of Feldspar embedded in a diabase ground mass. Well polished. 4 by 3½ ins.

Orezza Valley, Corsica.

No. 769. ANTIQUE GREEN PORPHYRY, or *Porfido verde antico.* Oriental Verd-antique. A porphyritic rock of the composition of dolerite. The Feldspar crystals (Labradorite) are comparatively large, with augite on a green base, deriving its colour from chlorite or viridite. Octagonal cut specimen, well polished. 3½ by 2½ ins. Delesse obtained from the compact base—

Silica...	53·55
Alumina	19·34
Ferrous Oxide	7·35
Manganese, Protoxide ...	0·85
Lime	8·02
Soda and Potash	7·93
Water	2·67
	99·71

Mount Taygesus, Greece.

This well-known rock is largely used in decorative art particularly for the border of tables, as well as being made into solid vases, principally by the French lapidaries.

No. 770. EGYPTIAN DIORYTE. A very dark green, spotted with white. A greenstone in part. Occurs at Mt. Dokhan, the "Porphyrites Mons" of Ptolemy. An analysis of a Dioryte yielded :

Silica	54·65
Alumina	15·72
Ferric Oxide ...	2·00
Ferrous Oxide ...	6·26
Manganese (a trace)	
Magnesia	5·91
Lime	7·83
Potash	3·79
Soda...	2·90
Water and Ignition ...	1·90
	100·96

Well polished. 4¼ by 3¼ ins.

Near Cairo, Egypt.

No. 771. VARIOLYTE. A variety of Euphotide, containing aphanitic concretionary spheroids of Saussurite. These concretions yielded, Delesse :

Silica	56·12
Alumina	17·40
Chromium Oxide ...	0·51
Ferrous Oxide ...	7·79
Magnesia	3·41
Lime...	8·74
Soda...	3·72
Potash	0·24
Ignition	1·93
	99·86

4¼ by 3 ins.

Orezza Valley, Corsica.

No. 772. MARBLE, brownish ground, with yellow stripes. Devonian limestone. Well polished. 2 by 1¾ ins.

Ness Rocks, Teignmouth.

No. 773. A SILICATED ROCK, enclosing Chalcedony. Rectangular. Polished on one side. 3⅝ by 2¾ ins.

Siberia.

OBSIDIAN. VOLCANIC GLASS.

FORMERLY considered a mineral, but not now included in the mineralogical works, being assigned to the volcanic rocks. Occurs massive, with a conchoidal fracture. Black, of various tints, with a greyish metallic lustre. Specific gravity of 2·34 to 2·39. Hardness 6·0 to 7·0.

Composition : A silicate of alumina, with potash, soda, and iron oxide, being really a glass produced by volcanic action. Volcanoes form, as it were, a cement for complete mountain ranges within their vicinity. Under the microscope Obsidian reveals radiating or hair-like clusters, termed *Microlites*, or microscopic minerals called *Trichites*. It was known to Pliny under the name of Obsian stone or Obsian glass, and he states that the Romans manufactured mirrors and gems from it. The Mexicans and Peruvians used it largely, vast quantities of cores, flakes, etc., being found among their ruins. The flakes, through their keen, sharp edges, it is recorded, were used as razors.

Obsidian.

No. 774. OBSIDIAN, from Obsius, who discovered it, according to Pliny, in Æthiopia. VOLCANIC GLASS. *Lapis Obsidianus* of Pliny. Black, with a peculiar greyish chatoyancy, sometimes causing it to be called Iridescent Obsidian. Rectangular. An analysis of this Mexican Obsidian yielded Vauquelin :

Silica	78·00
Alumina	10·00
Ferric Oxide ...	2·00
Oxide of Manganese	1·60
Potassium... ...	6·00
Lime	1·00
	98·60

3⅝ by 2⅜ ins.

Cerro de las Navasjas, Mexico.

Humboldt states that Cortez mentioned, when writing to the Emperor Charles V., that he observed razors made of Obsidian in use at Tenochittan. Specimens are sold sometimes by jewellers *as black* "*Cat's-eyes,*" after being cut *en cabochon* the chatoyancy appearing on the top.

No. 775. OBSIDIAN. VOLCANIC GLASS. Black, associated with brown Obsidian, taken at first for Jasper, and of a mahogany colour. A thin section, quite translucent, and of special interest, as it exhibits, by transmitted light, the flow of the lava in a circular direction when arrested by solidification. Well polished on both sides. 3¼ by 2¾ ins.

Yellowstone Park, U.S.A.

No. 776. OBSIDIAN. VOLCANIC GLASS. Black, with a greyish chatoyancy associated with brown Obsidian of a mahogany colour. The brown was mistaken at first by many for Jasper. A thin section similar to No. 775, exhibiting by transmitted light the flow of the lava, and the point at which it was arrested. Well polished on both sides. 3⅛ by 2½ ins.

Yellowstone Park, U.S.A.

No. 777. BALL OF OBSIDIAN or VOLCANIC GLASS. Black, with a greyish opalescent light. Diameter, 1¼ ins. Cut, drilled, and polished at Oberstein.

Volcano of Popocatapetl, Mexico.

Granite.

No. 778. GRANITE. Quartz, Feldspar, and Mica associated. Oval specimen. Partly polished. 2½ by 1⅞ in.

Near Cairo, Egypt.

No. 779 GRANITE containing Feldspar in unusually large crystals, associated with Quartz, Mica, and Magnesite. Well polished on one side. 4½ by 3⅛ ins.

Langbanshyttan District, Sweden.

A most distinct and beautiful granite, which, if found in any quantity, would be invaluable for decorative building purposes.

Names of celebrated collections from which examples were obtained, and which are in the "Derby" Collection.

The Collection of Minerals formed by His Imperial Highness the late Duke Nicholas of Leuchtenberg is well known. It is one of the finest in the world, and is kept at the Château de Stein in Bavaria. The author, since 1867 to His Highness' decease in Paris in 1890, has had the honour of making many exchanges with him, procuring in this manner the fine specimen of Siberian Blue Topaz, No. 706 as well as the Precious Beryls, Nos. 687, 688. His Highness was a distinguished scientific mineralogist.

His Excellency P. Kotschubey is a well-known Russian mineralogist, from whom the author obtained the Beryl, No. 689.

The Marquis de Chigi, from whom example No. 17 was procured, was an earnest Italian mineralogist, collecting specimens principally from the Island of Elba.

Dr. Joseph Walter Tayler, son of Admiral Tayler, was a Greenland explorer. He was sent by the Danish Government, in 1862, to superintend the mines in West Greenland, and to colonize the country—having made previously several voyages. He sailed in the Erik with the intention of cutting through an ice-field from west to east, but the expedition unfortunately failed. He afterwards went to assist to colonize "Prince Edward's Island," in the Gulf of St. Lawrence. The author knew Dr. Tayler well, and purchased the whole of his Minerals brought from the Greenland expedition, including the specimens represented in this Collection, Nos. 145 147, 148, 159, etc.

Dr. (now Sir George) Birdwood made a magnificent collection of Indian Agates in Bombay, purchasing for a great number of years every fine Agate brought in by the dealers from Central India. His collection was exhibited at South Kensington in 1871, attracting great attention.

Among the admirers was Mr. Arthur Wells, of Nottingham, who, upon their removal, sought Sir George Birdwood out and, *nolens volens*, insisted upon purchasing them. Sir George, being greatly pressed to name a price, mentioned one so high that he never dreamt would be accepted.

Mr. Arthur Wells, however, was not to be denied, and to Sir George's great astonishment and chagrin, wrote out the cheque for the sum mentioned, and took away the Agates.

Mr. Arthur Wells was a most distinguished and liberal collector of ornamental stone and jade ornaments. A magnificent collection of the latter he bequeathed to the South Kensington Museum, whilst the remainder, with his wonderful Agates, including those of Sir George Birdwood's, were dispersed by auction at Christie's—many being procured by the author and placed in the Derby Collection.

The collection of John Luff, Esq., the eminent Indian engineer, was made in various parts of India, and consist mostly of Mocha Stones; Nos. 382, 383, etc., belonged formerly to his collection.

The late Count Henri de Laurençel, of the Château de Pompadour, Fontainebleau, made a magnificent collection of Minerals, particularly those from France and South America. No. 7 and others were procured by exchange from him.

Prof. James Tennant was Lecturer upon Mineralogy for many years to King's College. The beautiful specimen of Moss Agate No. 367, was purchased from him by the author.

Specimens are also in the Collection from Prof. Louis Bombicci, the distinguished Mineralogist of Bologna; the late Mr. David Forbes, the eminent chemist; Captain Pulleine, a well-known conchologist as well as collector of minerals; Professor Charles Upham Shepard, of Amherst College, U.S.A.; Mr. A. M. Jacob, of Simla; His Excellency the late Julian de Siemaschko, and Herr Staats-rath Braun.

INDEX.

S

www.ingramcontent.com/pod-product-compliance
Lightning Source LLC
Chambersburg PA
CBHW021519210326
41599CB00012B/1309